Critical Introductions to Geography

Critical Introductions to Geography is a series of textbooks for undergraduate courses covering the key geographical subdisciplines and providing broad and introductory treatment with a critical edge. They are designed for the North American and international market and take a lively and engaging approach with a distinct geographical voice that distinguishes them from more traditional and outdated texts.

Prospective authors interested in the series should contact the series editor:
John Paul Jones III
Department of Geography and Regional Development
University of Arizona
jpjones@email.arizona.edu

Published

Cultural Geography: A Critical Introduction
Don Mitchell

Political Ecology: A Critical Introduction
Paul Robbins

Political Ecology

A Critical Introduction

Paul Robbins

BLACKWELL PUBLISHING
350 Main Street, Malden, MA 02148-5020, USA
9600 Garsington Road, Oxford OX4 2DQ, UK
550 Swanston Street, Carlton, Victoria 3053, Australia

First published 2004 by Blackwell Publishing Ltd

6 2006

Library of Congress Cataloging-in-Publication Data

Robbins, Paul, 1967–
Political ecology : a critical introduction / Paul Robbins.
p. cm. – (Blackwell critical introductions to geography)
ISBN 1-4051-0265-9 (hardback : alk. paper) – ISBN 1-4051-0266-7
(paperback : alk. paper)
1. Political ecology. I. Title. II. Series.

JA75.8.R63 2004
304.2–dc22

2003021513

ISBN-13: 978-1-4051-0265-0 (hardback : alk. paper) – ISBN-13: 978-1-4051-0266-7
(paperback : alk. paper)

A catalogue record for this title is available from the British Library.

Set in 10/12 Times
by Graphicraft Ltd, Hong Kong
Printed and bound in India
by Replika Press Pvt Ltd, Kundli

The publisher's policy is to use permanent paper from mills that operate a sustainable forestry policy, and which has been manufactured from pulp processed using acid-free and elementary chlorine-free practices. Furthermore, the publisher ensures that the text paper and cover board used have met acceptable environmental accreditation standards.

For further information on
Blackwell Publishing, visit our website:
www.blackwellpublishing.com

For

Vicki Robbins
Mari Jo Joiner

and

Martin Robbins

Contents

Figures

Tables

Boxes

Introduction

Driving through the Lamar Valley of Yellowstone last month, binoculars on the seat beside me and hoping to catch a glimpse of one of the now numerous wolves in the region, I could only imagine what the place might have looked like without the road that had conveniently brought me there. What might it have been like to walk to this place? Of course, even in the presence of the road, the view to the south, opening as it did across willow and aspen stands in the low drainages and upwards to the rocky slopes beyond, certainly *looked* like a wilderness.

Even so, if Yellowstone National Park is a wilderness, it had to be made into one through great political effort and expense. The history of the reserve, recognized as the crown jewel in the United States park system and a model for conservation everywhere, is in fact a testimony to human ingenuity, activity, and intervention.

Native American people had occupied and utilized the area in and around Yellowstone for ten thousand years prior to the park's establishment, clearing land through the use of fire and blazing trails across the landscape. Indeed, native hunting pressure probably served to concentrate the elk, antelope, and other animals that made the site so attractive to Anglo-Americans who later occupied the land. Native hunting pressure probably also allowed complex riparian ecosystems of aspen and willow to thrive. Removing these very human influences – those of the Shoshone, Bannock, Blackfoot, Flathead, Nez Perce, Utes, Crows, Piegans, and Paiutes who had helped to *produce* the very conditions later Americans would covet as "natural" – was an essential first step towards producing a wilderness.

This painful (and indeed violent) irony was not the only one. In much the same way, the establishment of the park coincided with the period in which the last gray wolf, prehistoric resident of the region, was shot dead. The absence of these predators, however, until their reintroduction in the mid-1990s, in no way kept most people from imagining Yellowstone as a "natural" system. Similarly in the late 1950s, when elk populations exploded, largely as a result of the absence of wolves, park personnel culled the herd, shipping animals to other ranges or simply shooting them. In the late 1960s outcry brought the practice to a halt, as the public was repelled by the apparently "unnatural" practices of the park's caretakers.

The banning of fire from the park, a dominant policy for many years, follows a similar history. This historically dominant practice gave way to management where natural fires were allowed to burn. The long suppression, followed by a dry season, led to the summer of fires in 1988 that still stands as a hallmark for "nature's wrath" in the popular press.

The removal and reintroduction of the wolves, the culling and protection of the elk, the abolition and return of fire, indeed the very establishment of a "wilderness" reserve from a sacred hunting and living space, are all human acts. They are, moreover, political ones. Each decision and counter-decision is born of bureaucratic incentives, economic pressures, and the changing power of rangers, legislators, hunters, concession companies, hoteliers, ranchers, visitors, environmentalists, and scientific experts in an ongoing struggle. Yellowstone is an expression of political power both in its very existence, as well as in the specific distribution of species across its landscapes. Nor are the political, economic, and ecological stakes in this struggle trivial. The livelihoods of tens of thousands of people, the power and budgets of large government agencies, the fate of rare and endangered species, and the environmental characteristics of tens of thousands of square miles all hang in the balance.

As I pulled the car over to the side of the road and stopped to watch the ridgeline, I kept thinking about all that *struggle* hidden behind the quiet vista. Yellowstone, it seemed, is a political project with contradictory mandates. The park is designed to produce elk for the regional hunting economy while producing wolves for scientists and environmentalists, and to produce open range for wildlife reproduction while producing vistas for visitors who travel thousands of miles to see bison and bear. No wonder its history is marked by a chaotic seesaw, which in one moment slaughters animals and the next moment protects them and in one instant suppresses fire and the next allows it. To be sure, in order to explain the current environment of Yellowstone, its species distribution, its forest cover, its water drainage patterns, there is simply no way to ignore the pounding political rhythms that thrum behind the visible backdrop of trees and snow-capped mountains.

But there is something more. The environment actually produced in this apparently endless struggle, the one I looked out on through the window of my rented SUV (sports utility vehicle), is by no means the one that any of the parties to the fight might have ever predicted. Burned landscapes are clear of over-mature lodgepole pines we typically associate with the area, and are dominated instead by brush and grasses in a surprising patchwork. The wolf population, now well in excess of 100 animals, includes individual packs larger than those seen for more than a century, defying the predictions of even the most optimistic biologists. Elk populations continue to fluctuate with effects on local vegetation that remain unclear, especially in light of the possible confounding factor of climate change.

So even amidst the countless cabins and hotels, the paved all-weather roads, and the concrete concessions that mark this very human landscape, complicated ecological interactions create a world of unintended consequences and surprises defying even the most careful political assessments or predictions. The system continues to stubbornly present new challenges for managers and new conditions for political wrangling. In a curious way, political actors create the ecology of Yellowstone, but not the Yellowstone of their own choosing.

This book is an effort to address that tension. By introducing political ecology, a field that seeks to unravel the political forces at work in environmental access, management, and transformation, I hope to demonstrate the way that politics are inevitably ecological and that ecology is inherently political. But more than this, I intend to show that research in the field can shed light on environmental surprise and dynamism, like that evident at Yellowstone, so addressing not only the practical problems of equity and sustainability, but also basic questions in environmental science.

The normative goal of the book is not over-ambitious. By explaining and constructively exploring the work of a group of researchers sometimes called political ecologists, I intend only to clarify the most persuasive themes in a highly disparate body of writing and show the politics of nature to be both universal and immediate. This, I think, may make a small contribution to helping us all break from an image of a world where the human and the non-human are disconnected, a fiction that remains so stubborn a part of our modern reasoning that it is as difficult to unimagine as it is to picture a world without patriarchy or class. I believe, however, that an alternative picture, where nature and society are undivided, is as much an act of remembering as one of inventing. Since the popular environmental movement has already done such an admirable job of getting many of us started, it may only be a matter of completing the revolution by rendering it more explicitly political.

It is my hope, therefore, that though this book is aimed at an academic audience, it presents the claims of the field in a plain enough way that national park visitors, whale watchers, and birders can find in it a compelling argument for the way their concerns are implicated in those of working communities, disenfranchised minorities, and subsistence producers around the world. In this sense the book departs from some theoretical and programmatic approaches to the politics of nature, especially those that eschew alliances with traditional environmental movements. This rejection of "bourgeois" environmentalism, a hallmark of some political economic approaches to nature, is both shortsighted and impractical; what more radical challenges to the political economic *status quo* exist in US law than the Clean Air Act, the Clean Water Act, or the Endangered Species Act?

Having said this, it is also my goal to persuade those concerned about the fate of forests, dolphins, and wolves that it is no blasphemy to admit the world to be crafted by political forces and human industry, even and especially those dearly held wildernesses that sell so many Sierra Club calendars. At the same time I hope to encourage those concerned with more traditional political economy that an increased sensitivity to the influence (and perhaps even the interests) of non-humans is essential for better politics, explanation, and ethics. The potential power of a popularized political ecology is so great, in fact, that merely shedding a few tightly clasped shibboleths on either side might make way for a very new world, emerging from these dark times when progressive politics in both human and non-human realms seem so painfully paralyzed.

The Goals of the Text

It would be impossible to survey the field of political ecology in its entirety. The contributors are too many, the breadth of topics too vast, and the regional diversity too great. I do not, therefore, intend here to provide exhaustive case studies of political ecological research (see especially Peet and Watts 1996a) or a general account of the relationship between science and politics (Forsyth 2003), since this is a task well performed by others. Neither do I intend to survey the world system as a whole, pointing to the processes, players, and dynamics that are at work politicizing the natural environment. Many excellent books survey the condition of global debt, the position of local producers in commodity markets, and the dwindling power of the state in managing nature (see especially Porter and Sheppard 1998 and Bryant and Bailey 1997).

Rather, I intend to do something different here. Whereas most summary texts on the state of global political ecology are designed to show political ecology as a *body of knowledge*, this book is designed also to show political ecology as *something people do*. And whereas collected volumes highlight a number of separate and *distinct cases*, this book also gropes for *common questions* that underlie them.

The book is designed to serve as an introduction and companion volume to the key books, articles, arguments, and research statements that make up the core of the field, and should serve to introduce any interested party to its major works and contested ideas. In this regard, it is offered as a remedy for the purported problem that the field is so fragmented that citation in it, as senior political ecologist Piers Blaikie once remarked, "is largely a random affair."

But more than this, the book is a critical review of the work that goes on in the field, one that advocates a very particular vision of which approaches work and which do not and which lines of inquiry have the most political and analytic power and which do not. In the process, I further hope that the book reveals three areas where the field might yet improve its analytical tools. First, political ecology would benefit from a shift away from a view of environmental change in terms of either the "destruction" of nature or its social "construction" and towards a view of the *production* of nature by human and non-human actors, with varying (often serious) normative implications. Second, political ecologists would benefit from a broader examination of *all* producers of nature, including ministry chiefs, SUV drivers, forestry professionals, as well as the herders, farmers, and wood-cutters of traditional concern. Third, the field will improve with a departure from linear "chains" of explanation towards "networks," with an exploration of the complex and shifting connections that this implies. Even while showing the strength of the approach, therefore, the book is written to demonstrate weaknesses, while pointing the way forward towards a more coherent and simultaneously more critical way of doing research.

I will not provide and rehearse, however, the laundry list of more typically pronounced criticisms often made of the field – usually centered on the fact that it is too focused on the broadly defined "underdeveloped world." This is true, but such biases, as discussed here, grow quite inevitably from the professional and intellectual seeds from which the tree of political ecology sprouted – critical development research, peasant studies, environmental history, cultural ecology, and postcolonial theory. In the future, I am confident, political ecology will become more concerned with the traditionally defined "first world." This change will not guarantee, however, that its approaches will be more coherent, or that the use of either ecological science or critical deconstruction will be managed with greater rigor. These problems, I argue, are prior to and more important than the specific topical and regional choices made in research. Their resolution remains vital to the continued growth and effectiveness of the field, both as an intellectual investigation of the human–environment interaction, and as a political exercise for greater social and ecological justice.

The Rest of the Book

The remainder of this book directs itself to describing political ecology as a way to do research, attempting to show what makes political ecology researchers tick, what makes their work urgent to them, and what useful lessons they have provided for addressing the important questions of environmental change around the world. More than this, I intend to show the yet untapped possibilities for the field and the possible vistas onto which we might yet gaze using this approach.

In part I, I describe how political ecology came to be the way it is, with its inherent possibilities and limits. Chapter 1 introduces the term political ecology and its many definitions, showing a unity of practice amidst much diversity of thought. Chapter 2 reviews the deep roots of this line of inquiry, arguing that political ecologists have been around a very long time. Chapter 3 describes the historical development of a critical science of the environment, showing the disparate fields and eclectic range of tools that converged in the last three decades of the twentieth century to give greater analytical form to the field. This chapter is dense with history and referencing, but is intended to be a source to which the reader can return while exploring the rest of the volume, which depends heavily on the concepts introduced there. Chapter 4 describes the crystallizing period of political ecology in the 1980s, with case examples from the Sahelian crisis to demonstrate the key concepts of chains of explanation and situated behaviors.

Part II reviews the ways in which political ecologists commonly address the environment, either in terms of "land degradation" or "social construction," and the methodological and conceptual problems facing either inquiry. Chapter 5 examines the question of environmental change as environmental degradation or destruction, while Chapter 6 attends to the way researchers have considered the environment to be imaginary or constructed.

Part III examines four central theses of political ecological research, each in its own chapter, which I describe as (1) degradation and marginalization, (2) conservation and control, (3) environmental conflict, and (4) environmental identity and social movement. The case materials in each chapter are selected to represent a range of research regions across the world, including cases from the "developed" and "underdeveloped" worlds.

The biases of my training will be evident throughout this section. The research described comes predominantly from the discipline of geography, though it is coupled with work in environmental history, development studies, anthropology, and sociology. While I have tried to include examples from both the global north and south, including cases from North and South America, Africa, and Asia, I have mentioned nothing of Western or Eastern Europe or of Australia. Research and theory in English predominates in the volume, despite the strong parallel threads of Francophone political ecology (Whiteside 2002). Referencing of North American work somewhat outweighs that from other places, largely again as a result of my own training. Finally, numerous international case examples were cut in final editing, owing to a lack of space.

Each of the chapters in this section also includes case histories of how, in my own work, I have tried to do research, and how on many occasions I have been tripped up by the hidden pitfalls of fieldwork. These sections only reflect what I have done in research rather than what political ecologists have done more generally, but I think my methodological choices are not unique and the problems I have faced are common not only to political ecology, but to much research in general.

The conclusions in part IV will critically evaluate the status of the field and point to both topical and conceptual areas where political ecology can expand and improve. Conceptually, I urge a movement from studies of environmental "destruction" to "production," from "peasants" to "producers," and from "chains" to "networks" of explanation. Substantively, I point to areas where too little political ecology is done, stressing that: (1) population is too important to be left to the Malthusians, (2) genetically modified landscapes won't go away and will be a part of global ecologies very soon, and (3) cities are subsistence ecologies. These collective lessons, I further suggest, constitute the fifth and most recent thesis in political ecology: the hybridity thesis.

Scattered throughout the text are boxed critical summaries of important individual contributions to political ecology and the people who made them. These are based on my own reading, but wherever possible these also include direct reflections and responses from those authors kind enough to provide them.

Many Acknowledgments

I am not a political ecologist, though I've often tried to be one. As will be evident from the content and organization of the book, which journeys far beyond my own limited global experiences, constrained theoretical understandings, and finite grasp of politics, economics, and ecology, this work could not have been finished without *a lot* of help. First, all of the researchers I approached in the preparation of the volume, and whom I asked to share their own perspectives and feelings about their work and the history of the field, were invaluable, including Tom Bassett, Fikret Berkes, Piers Blaikie, Harold Brookfield, Judith Carney, Larry Grossman, Nancy Peluso, Diane Rocheleau, and Michael Watts. Second, I am in debt to my many colleagues around the world, who answered endless emails, sent photographs, edited drafts, and explained complex problems so that even I could grasp them, including Simon Batterbury, Tony Bebbington, Susanna Hecht, Noriko Ishiyama, Brad Jokisch, Thembela Kepe, Rheyna Laney, Becky Mansfield, Kendra McSweeney, Ian Scoones, and Randy Wilson. John Isom provided great feedback on early drafts and we together produced figures 2.2 and 3.1.

The several years of my own fieldwork described throughout the book would have been impossible without the ongoing help of Doug Johnson, Jody Emel, Ilse Kohler-Rollefson, Hanwant Singh Rathore, S. M. Mohnot, and Komal Kothari. Thanks also go out to John Paul Jones, whose idea this was in the first place, and to the folks at Blackwell, Sarah Falkus, Justin Vaughan, and Simon Alexander. Thanks to Bob

Toborg, Frank Forgione, and Dave Feroe for being the (extremely demanding) intended audience of the volume. Special thanks go to Dr John J. Sharkey at Ohio State University orthopaedics, who inserted titanium into my ankle expertly enough to allow me to hobble to the office. Finally, and very likely against his wishes, Billie Lee Turner II remains both an inspiration and a foil for this work.

Most importantly, throughout the whole process Sarah Moore continued to insist not only that the book would eventually get finished (despite my strong doubts) but that at least one person would eventually agree to read it; her comments on and support for my writing have saved a great many confusions and embarrassments over the years (the word "penultimate" means next to last, for example; who knew?). Having said this, the interpretations and perspectives contained within the book are my own, and I certainly can't lay blame at anyone else's feet for controversial, confusing, or bizarre claims. The reader will have to address any complaints to me.

Part I

What is Political Ecology?

In which eclectic uses of the term "political ecology" are introduced and wherein much divergent research is shown to share some important common questions. Rather than finding a single body of theory, we discover instead a number of independent trains of thought colliding in the field, leading to a remarkable synthesis in the late 1980s.

1

The Hatchet and the Seed

For many of us who are unable to travel to the plains of East Africa, our images of the region are given life on late-night cable wildlife television, in bold IMAX presentations at natural history museums, or perhaps in the vivid spectacle of Disney's *The Lion King*. The imagined patterns of the "circle of life" in these media – complete with lions, hyenas, and baboons – play out on a yellow-filtered savanna where migrations of wildebeest cross the Serengeti chasing seasonal rainfall, hunted in turn by stoic predators. The scenes are compelling and they inspire in us a justifiable affection for the beauty and complexity of the non-human world around us. These images are also ecologically important, since they give us a picture of connectedness, which is essential to understanding life on the savanna. Across the borderlands of Kenya and Tanzania forage grasses follow rainfall, wildebeest pursue forage, predators pursue wildebeest, scavengers pursue predators, and so on.

The absence of people from these imaginary landscapes seems in no way strange for most of us; these are *natural* landscapes, apparently far from farms, factories, and the depredations of humankind. It is perhaps inevitable, therefore, that an intuitive reaction to the news that wildlife populations are in crisis – including declines in giraffe, topi, buffalo, warthog, gazelle, and eland – is to imagine that the intrusion of humankind into the system is the cause of the problem. Growing populations of impoverished African people, we might imagine, have contaminated the natural rhythm of the wilderness. Indeed, the sense of loss in contemplating the declining biodiversity and destroyed landscapes may inspire frustration, coupled with a feeling of helplessness; the situation in the Serengeti and the steady march of growing populations seem far beyond the control and influence of life where we live.

Stepping back from the savanna, however, and gazing across Serengeti–Mara ecosystem both in time and space, habitat loss and wildlife decline appear both more complex and more connected to the daily lives and routines of urban people in the developed world. Cross-border analysis shows the decline in habitat and wildlife in Kenya is far higher than in Tanzania. Why? Rainfall, human population, and livestock numbers do not differ significantly. Rather, private holdings and investment in export cereal grains on the Kenyan side of the border have led to intensive cropping and the decline of habitat. These cereals are consumed around the world, as part of an increasingly globalized food economy. As Kenya is increasingly linked to these global markets and as pressure on local producers increases, habitat loss is accelerated. Less developed agricultural markets and less fully privatized land tenure systems in Tanzania mean less pressure on wildlife. The wildlife crisis in East Africa is more political and economic than demographic (Homewood et al. 2001).

These facts undermine widely held apolitical views about ecological relations in one of the most high-profile wildlife habitats in the world. They also point to faulty assumptions about the nature of "wild" Africa. Firstly, the image of a Serengeti without people is a fallacious one. The Massai people and their ancestors inhabited the Central Rift Valley for thousands of years before European contact, living in and around wildlife for generations. Indeed, their removal from wildlife park areas has led to violent conflicts (Collett 1987). More generally, the isolation of these places is also a mistaken perception. Export crops from Kenya, including tea and coffee in other parts of Kenya beyond the Central Rift Valley, continue to find their way to consumers in the first world, even as their global prices fall, constraining producers who must increase production, planting more often and over greater areas, further changing local ecological conditions. With three-quarters of the population in agriculture, economic margins for most Kenyans become tighter every year, and implications for habitat and wildlife more urgent.

The migration of the wildebeest, and its concomitant implications for grasslands and lions, therefore, does not occur outside the influences of a broader political economy. Land tenure laws, which set the terms for land conversion and cash cropping, are made by the Kenyan and Tanzanian state. Commodity markets, which determine prices for Kenyan products and the ever-decreasing margins that drive decisions to cut trees or plant crops, are set on global markets. Money and pressure for wildlife enclosure, which fund the removal of native populations from the land, continue to come largely from multilateral institutions and first-world environmentalists. All of these spheres of activity are further arranged along linked axes of money, influence, and control. They are part of systems of power and influence that, unlike the imagined steady march of the population "explosion," are *tractable to challenge and reform*. They can be fixed.

The difference between this contextual approach and the more traditional way of viewing problems like this is the difference between a *political* and an *apolitical* ecology. This is the difference between identifying broader systems rather than blaming proximate and local forces; between viewing ecological systems as power-laden rather than politically inert; and between taking an explicitly normative approach rather than one that claims the objectivity of disinterest.

When the bottom drops out of the coffee market, as it did in the late summer of 2001, what happens to the peasants who depend upon it and the forests in which it is harvested? When the World Bank helps to fund massive afforestation programs around the world, aimed at preserving tree cover and animal biodiversity, what actually happens to the hill forests designated for enclosure and the tribal people who live there?

These are the questions of political ecology, a field of critical research predicated on the assumption that any tug on the strands of the global web of human–environment linkages reverberates throughout the system as a whole. This burgeoning field has attracted several generations of scholars from the fields of anthropology, forestry, development studies, environmental sociology, environmental history, and geography. Its countless practitioners all query the relationship between economics, politics, and nature but come from varying backgrounds and training. Some are physical scientists (e.g., biologists, geomorphologists, and hydrologists), others are methodological technicians (e.g., geographic information or remote sensing specialists), while most are social and behavioral scientists, who share an interest in the condition of the environment and the people who live and work within it. These researchers, moreover, advocate fundamental changes in the management of nature and the rights of people, directly or indirectly working with state and non-governmental organizations to challenge current conditions. This book reviews the work that these people do, pointing towards the common factors evident in a research area often noted for its diversity, and revealing the strengths and weaknesses in a field that has grown far too quickly to prepare a comprehensive survey or census of its accomplishments and failures.

What is Political Ecology?

The term political ecology is a generous one that embraces a range of definitions. A review of the term from its early use (probably first coined by Wolf in 1972) to its most recent manifestations shows important differences in emphasis. Some definitions stress political economy while others point to more formal political institutions; some identify environmental change as most important, while others emphasize narratives or stories about that change (see list below). Even so, there seems to be a set of common elements. The many definitions together suggest that political ecology represents an explicit alternative to "apolitical" ecology, that it works from a common set of assumptions, and that it employs a reasonably consistent mode of explanation.

Defining Political Ecology

Author/Source	*Definition of "political ecology"*	*Goal*
Cockburn and Ridgeway (1979)	"a useful way of describing the intentions of radical movements in the United States, in Western Europe and in other advanced industrial countries . . . very distant from the original rather sedate operations of the eco-lobby" (p. 3)	Explicate and describe first world urban and rural environmental degradation from corporate and state mismanagement; document social activism in response
Blaikie and Brookfield (1987)	"combines the concerns of ecology and a broadly defined political economy. Together this encompasses the constantly shifting dialectic between society and land-based resources, and also within classes and groups within society itself" (p. 17)	Explain environmental change in terms of constrained local and regional production choices within global political economic forces, largely within a third world and rural context
Greenberg and Park (1994)	A synthesis of "political economy, with its insistence on the need to link the distribution of power with productive activity and ecological analysis, with its broader vision of bio-environmental relationships" (p. 1)	"Synthesize the central questions asked by the social sciences about the relations between human society, viewed in its bio-cultural-political complexity, and a significantly humanized nature" (p. 1)
Peet and Watts (1996b)	"a confluence between ecologically rooted social science and the principles of political economy" (p. 6)	Locates "movements emerging from the tensions and contradictions of under-production crises, understands the imaginary basis of their oppositions and visions for a better life and the discursive character of their politics, and sees the possibilities for broadening environmental issues into a movement for livelihood entitlements, and social justice" (pp. 38–9)
Hempel (1996)	"the study of interdependence among political units and of interrelationships between political units and their environment . . . concerned with the political consequences of environmental change" (p. 150)	Explore and explain community-level and regional political action in the global sphere, in response to local and regional degradation and scarcity
Watts (2000)	"to understand the complex relations between nature and	Explain environmental conflict especially in terms of struggles over

	society through a careful analysis of what one might call the forms of access and control over resources and their implications for environmental health and sustainable livelihoods" (p. 257)	"knowledge, power and practice" and "politics, justice and governance"
Stott and Sullivan (2000)	"identified the political circumstances that forced people into activities which caused environmental degradation in the absence of alternative possibilities . . . involved the query and reframing of accepted environmental narratives, particularly those directed via international environment and development discourses" (p. 4)	"Illustrating the political dimensions of environmental narratives and in deconstructing particular narratives to suggest that accepted ideas of degradation and deterioration may not be simple linear trends that tend to predominate" (p. 5)

Challenging apolitical ecologies

If there is a political ecology, by implication there must exist an apolitical one. As such, research in the field commonly presents its accounts, whether explaining land degradation, local resource conflict, or state conservation failures, as an alternative to other perspectives. The most prominent of these apolitical approaches, which tend to dominate in global conversations surrounding the environment, are "ecoscarcity" and "modernization" accounts.

It is not my intention to provide sustained criticisms of these two approaches here; later chapters of the book should reveal the characteristics of these perspectives and demonstrate their ethical and practical weaknesses. An outline of each should suffice to present their basic arguments, with which readers are probably already very familiar, common as these approaches are to most environmental explanation.

Ecoscarcity and the limits to growth The dominant contemporary narrative of environmental change and human–environment interaction is a well-established one with a long history. In Western Europe since the late 1700s, when human influence and response to the environment was first submitted to scientific scrutiny, the central driving explanation for social/ecological crisis has been increasing human population, measured in absolute numbers. Following from Thomas Malthus's *Essay on the Principle of Population*, the argument is straightforward: as human populations grow out of proportion to the capacity of the environmental system to support them, there is a crisis both for humans, whose numbers fall through starvation and disease-based mortality, and for nature, whose overused assets are driven past the point of self-renewal. This argument took many forms during the twentieth century, from the "population bomb" of Paul Ehrlich (1968) to the Club of Rome's "Limits to Growth" (Meadows et al. 1972), but its elements are consistent. All hold to the ultimate scarcity of non-human nature and the rapacity of humankind's growing numbers.

Table 1.1 Who is overpopulated? Comparative per capita consumption of resources

Resource	*India*	*United States*
Energy (kg oil equivalent)	477	7,956
Meat (kg)	4	122
Paper (kg)	4	293
Water (m^3)	588	1,844

For ecoscarcity proponents, this is nowhere a more serious problem than in the underdeveloped world, where growth rates and absolute numbers of people remain the highest in the world. That the poorest regions of the world are the repositories for what are viewed as important and scarce environmental goods makes the problem doubly serious. In this way of thinking, the perilous decline of Kenya's wildlife, as described above, can be predicted to follow inevitably from the growth of Kenya's population.

The problems with this line of argument are many. In general terms, and as will be shown throughout this book, the demographic explanation is a consistently weak predictor of environmental crisis and change. Firstly, this is because the mitigating factors of affluence and technology (following Commoner 1988) tend to overwhelm the force of crude numbers. A very few members of the global village consume the majority of its resources. When these factors are considered, overpopulation, to the extent that such a thing exists on a global or regional scale, appears to be a problem strictly of smaller, wealthier populations, especially the United States, rather than the apparently larger populations of the global south (Table 1.1).

The more fundamental problem with this formulation, however, is that it posits the environment as a finite source of basic unchanging and essential elements, which set absolute limits for human action. However intuitive (divide a limited stock of earth materials by a potentially infinite hungry human population and the result is zero), this assumption has proved historically false and conceptually flawed.

Market "optimists," expressing the problem in economic terms, suggest that any form of resource scarcity creates a response that averts serious crisis. As a good becomes scarcer, they suggest, its price tends to rise, which results either in the clever use of substitutes and new technologies to increase efficiency, or in a simple decreased demand for that good. The result is that apparently finite resources are stretched to become infinitely available as consumers use less and producers supply more efficient alternatives and substitutes (Rees 1990). Even if populations rise on a limited land area, for example, the demand for land and rising land rents will increase its efficiency of use, with more and better production on each unit of land. Even if petroleum becomes scarce, the rising price per barrel will encourage the use of otherwise expensive alternatives like wind and solar power, or simply cause consumers to drive less, endlessly stretching the world's energy supply. While such optimistic prognoses are themselves fraught with problems, they do point to an important and increasingly well accepted truism: resources are constructed rather than given.

This is not to argue that the number of organisms versus the extent and character of local resources is not an important issue; ask anyone who is in charge of extending water services to suburbs outside of Denver or Phoenix. To be sure, the number of people who use trees, food, water, metals, and other materials in part determines proximate demands on the environment. So too, the adaptation of natural systems to meet changing needs, whether driven by absolute numbers or changing consumption patterns, is an important element of human–environment interactions. Even so, the Malthusian population pressure model poorly reflects the complexity of global ecology. The argument does, however, hold serious implications for the use and management of resources.

When it was first offered up in Malthus's 1793 formulation, the ecoscarcity argument was presented as an explicit justification for social policy. Specially, Malthus insisted that since famine and starvation were essential to controlling runaway human populations, such events are "natural" and inevitable. England's Poor Laws, the modest redistributive welfare subsidies to feed the most marginal groups, were pointless and counter-environmental. By increasing rather than decreasing their numbers, such subsidies were the source not the solution of misery. So too, in such a conceptualization, the crisis for the poor lay not in the larger economy or ecology of their subsistence, but instead in and amongst the poor themselves: "In searching for objects of accusation, [the poor man] never adverts to the quarter from which all his misfortunes originate. The last person he would think of accusing is himself, on whom, in fact, the whole blame lies" (Malthus 1992, book 4, chapter 3, p. 227).

The implications for contemporary global environmentalism are equally programmatic. Environmental crises as demographic problems exist at the site of resource use, in and amongst the world's poor, who are simply too numerous. Subsidies of the poor do little to alleviate the crisis, since they only serve to reinforce the demographic trend. Population control, rather than reconfiguration of global distributions of power and goods, is the solution to ecological crisis. The continued advocacy of an apolitical natural limits argument, therefore, is implicitly *political*, since it holds implications for the distribution and control of resources.

Demographic explanations of environmental change have become considerably more sophisticated than those outlined by Malthus and the Club of Rome. Attention to high-density urban development and the associated energy costs and infrastructure demands of mega-cities have created justifiably renewed attention to population as an important driver for environmental change. More recent research has come to demonstrate that the position of women in the workforce and their increased access to decision-making, calories, and education are closely linked not only to changing environmental conditions but also to decreased fertility and population growth. New approaches have come to redefine our ways of thinking about population, power, and environment. Even so, crude Malthusianism regrettably remains a typical way of thinking about environmental change, and so provides a unifying target for many political ecologists.

Other apolitical ecologies: diffusion, valuation, and modernization Other prominent accounts of environmental change also dominate current thinking, asserting apolitical answers to extremely political questions. It is commonly argued, for example, that ecological problems and crises throughout the world are the result of inadequate adoption and implementation of "modern" economic techniques of management, exploitation, and conservation. Generally, this way of thinking is underpinned by a commitment to economic efficiency.

These approaches to environmental management and ecological change generally assert that efficient solutions, determined in optimal economic terms, can create "win-win" outcomes where economic growth (sometimes termed "development") can occur alongside environmental conservation, simply by getting the prices and techniques right. Such approaches are persuasive, at least insofar as they reject the cataclysmic prognoses of Malthusian catastrophe described above. The assertion that economic efficiency pays environmental dividends is further supported by many examples over the recent period of industrial technological change. The historically dirty pulp and paper industry, in a prominent example, has simultaneously increased profit margins and decreased emissions through efficient industrial ecological practices (Pento 1999). By freeing individuals and firms to seek their own best and most efficient use of resources, propelled by competition on an open market and sustained by modern technology, waste, environmental destruction, and resource degradation can be tamed. Moreover, the sometimes perverse influence of strong state bureaucracies over the environment are perhaps avoided through market- and technology-based solutions.

For global ecology, such an approach suggests several general principles and policies. (1) Western/northern technology and techniques need to be diffused outwards to the underdeveloped world. (2) Firms and individuals must be connected to larger markets and given more exclusive property controls over environmental resources (e.g., land, air, wildlife). (3) For wilderness and biodiversity conservation, the benefits of these efficiencies must be realized through institutionalizing some form of valuation; environmental goods like wildebeest, air, and stream quality must be properly priced on an open market.

The debates and critiques surrounding such approaches and the logics that underpin them are too numerous to summarize here; even so, there are some serious general conceptual and empirical problems with this perspective. First, the assertion that modern technologies and markets can optimize production in the underdeveloped world, leading to conservation and environmental benefits, has proven historically questionable. The experience of the green revolution, where technologies of production developed in America and Europe were distributed and subsidized for agrarian production around the world, led to what even its advocates admit to be extensive environmental problems: exhausted soils, contaminated water, increased pest invasions (Lal et al. 2002). Beyond these failings, the more general assertion that superior environmental knowledge originates in the global north for transfer to the global south is in itself problematic, reproducing as it does paternalistic colonial knowledge relations and a priori discounting the environmental practices of indigenous and local communities (Uphoff 1988).

Articulation with global markets, as will be shown in the case materials presented here, has also proved to be a mixed environmental blessing at best. Changes in markets, falling commodity prices, and altered land values that have followed from globalized exchange have often led to land degradation and social disorder in the less developed world. A call to intensify these forms of exchange must be viewed skeptically. More generally, even in free and open markets, monopoly control of resources commonly perverts allocation and distribution, leading to far from optimal social and ecological outcomes. Indeed, the tradition of conservation in the United States is largely based on the understanding that collective control of environmental resources is necessary for fair and sustainable distribution.

Asserting and adopting the apparently apolitical approach to the environment suggested in market and modernization approaches, because of the institutional and political changes that such an approach mandates, is also inherently political. To individuate and distribute "collective" goods like forests or water by necessity requires the alienation of previous user groups. To implement new technological approaches in agriculture, resource extraction, or wilderness management requires a transformation of existing institutions. Increasingly open markets demand deregulation of labor and environmental controls. There is nothing apolitical about such a proposal.

The first lesson to draw is that the dominant contemporary accounts of environmental crisis and ecological change (ecoscarcity and modernization) tend to ignore the significant influence of political economic forces. As we shall see, this is to ignore the most fundamental problems in contemporary ecology. The other lesson is that apolitical ecologies, regardless of claims to even-handed objectivity, are implicitly political. It is not so much that political ecology is "more political" than these other approaches to the environment. Rather it is simply more *explicit* in its normative goals and more outspoken about the assumptions from which its research is conducted.

Common assumptions and modes of explanation

Following Bryant and Bailey, political ecological accounts and research efforts also share a common premise, that environmental change and ecological conditions are the product of political process. This includes three fundamental linked assumptions in approaching any research problem. Political ecologists: "accept the idea that costs and benefits associated with environmental change are for the most part distributed among actors unequally . . . [which inevitably] reinforces or reduces existing social and economic inequalities . . . [which holds] political implications in terms of the altered power of actors in relation to other actors" (Bryant and Bailey 1997, pp. 28–9).

Research tends to reveal winners and losers, hidden costs, and the differential power that produces social and environmental outcomes. As a result, political ecological research proceeds from central questions, such as: What causes regional forest loss? Who benefits from wildlife conservation efforts and who loses? What political movements have grown from local land use transitions?

In answering, political ecologists follow a mode of explanation that evaluates the influence of variables acting at a number of scales, each nested within another, with local decisions influenced by regional polices, which are in turn directed by global politics and economics. Research pursues decisions at many levels, from the very local, where individual land managers make complex decisions about cutting trees, plowing fields, buying pesticides, and hiring labor, to the international, where multilateral lending agencies shift their multi-billion dollar priorities from building dams to planting trees or farming fish. Such explanation also tends to be highly (sometimes recklessly) integrative. Bryant (1999) recently described the field as a series of "disciplinary transgressions" where researchers trace their personal and professional trajectories from political studies and sociology to geography or from geography to development studies.

So, at the risk of adding yet another definition to a crowded field, these many understandings of political ecology together appear to describe: empirical, research-based explorations to explain linkages in the condition and change of social/environmental systems, with explicit consideration of relations of power. Political ecology, moreover, explores these social and environmental changes with a normative understanding that there are very likely better, less coercive, less exploitative, and more sustainable ways of doing things. The research is directed at finding causes rather than symptoms of problems, including starvation, soil erosion, landlessness, biodiversity decline, human health crises, and the more general and pernicious conditions where some social actors exploit other people and environments for limited gain at collective cost. Finally, it is a field that stresses not only that ecological systems are political, but also that our very ideas about them are further delimited and directed through political and economic process. As a result, political ecology presents a Jekyll and Hyde persona, which attempts to do two things at once: critically explaining what is wrong with dominant accounts of environmental change, while at the same time exploring alternatives, adaptations, and creative human action in the face of mismanagement and exploitation.

The hatchet: political ecology as critique

As critique, political ecology seeks to expose flaws in dominant approaches to the environment favored by corporate, state, and international authorities, working to demonstrate the undesirable impacts of policies and market conditions, especially from the point of view of local people, marginal groups, and vulnerable populations. It works to "denaturalize" certain social and environmental conditions, showing them to be the contingent outcomes of power, and not inevitable. As critical historiography, deconstruction, and myth-busting research, political ecology is a hatchet, cutting and pruning away the stories, methods, and policies that create pernicious social and environmental outcomes.

These critical efforts have more recently been extended to encompass research that not only demonstrates the way many dominant accounts are wrong, but shows, moreover, how those accounts themselves are instrumental in political and ecological change. To take but one example, "The Pristine Myth" of the Americas, a story which holds that the landscapes of the New World were in an Edenic and "natural" order unaffected by human activity prior to European arrival, has been placed under political ecological scrutiny. As geographer William Denevan has demonstrated (1992, 2001), summarizing 30 years of his own archaeological, field, and historical research, pre-Columbian environments were heavily influenced by native peoples' cutting, planting, terracing, and building. Political ecology suggests, moreover, that the myth of a "pristine" environment was itself important in the colonial process of marginalizing and disenfranchising native peoples. By writing indigenous people out of the landscape, the business of control was easier to carry out (Sluyter 1999). Political ecology takes a hatchet to such stories.

The seed: political ecology as equity and sustainability research

This research has another side, which seeks to document the way individuals cope with change, households organize for survival, and groups unite for collective action. In this sense, much political ecology involves the detailed analysis of agrarian practices, social systems for resource distribution, and techniques for cataloging and harvesting non-human nature. Often this means careful attention to "traditional" ways things were done historically, documenting local knowledges and understandings of ecological process.

As Peet and Watts insist, however, this "concern is not simply a salvage operation – recovering disappearing knowledges and management practices – but rather a better understanding both of the regulatory systems in which they inhere . . . and the conditions under which knowledges and practices become part of alternative development strategies" (Peet and Watts 1996a, p. 11). In other words, political ecology seeks not simply to be retrospective or reactive, but to be progressive. A political ecological analysis of the decline of traditional water harvesting techniques under the increasing influence of state irrigation authorities, for example, is not simply a mournful or romantic call for a lost technological past. By documenting not only the changing economic and bureaucratic pressures under which water management is currently being transformed, but also detailing the way it is managed traditionally and describing techniques of local adaptation and resistance, political ecological research helps to plant the seeds for reclaiming and asserting alternative ways of managing water (Rosin 1993). The goal of any such effort is preserving and developing specific, manageable, and appropriate ways to make a living.

The Dominant Narratives of Political Ecology

In this sense, political ecology is something that people *do*, a research effort to expose the forces at work in ecological struggle and document livelihood alternatives in the face of change. This does not mean that political ecology is something that people do all the time. Much of this work is carried out by people who might never refer to themselves as political ecologists, or who might do so in only one sphere of their work. Neither is political ecology restricted to academics from the "first world." Indeed, the ongoing, small-scale, empirical research projects conducted by countless non-governmental organizations (NGOs) and advocacy groups around the world, surveying the changing fortunes of local people and the landscapes in which they live, probably comprise the largest share of work in political ecology. Published only in local meeting and development reports, this work is as much a part of the field as the well-circulated books or refereed journal articles of formal science.

Big questions and theses

What unites the diverse work in these many locations is a general interest in four big questions, themes, or narratives of research. Oversimply, political ecology research

Four theses of political ecology and the things they attempt to explain

Thesis	*What is explained?*	*Relevance*
Degradation and marginalization	Environmental change: why and how?	Land degradation, long blamed on marginal people, is put in its larger political and economic context
Environmental conflict	Environmental access: who and why?	Environmental conflicts are shown to be part of larger gendered, classed, and raced struggles and vice versa
Conservation and control	Conservation failures and political/economic exclusion: why and how?	Usually viewed as benign, efforts at environmental conservation are shown to have pernicious effects, and sometimes fail as a result
Environmental identity and social movement	Social upheaval: who, where, and how?	Political and social struggles are shown to be linked to basic issues of livelihood and environmental protection

has demonstrated (or attempted to demonstrate) the following four general theses, summarized above, each of which receives a chapter later in this volume.

The degradation and marginalization thesis Otherwise environmentally innocuous local production systems undergo transition to overexploitation of natural resources on which they depend as a response to state development intervention and/or increasing integration in regional and global markets. This may lead to increasing poverty and, cyclically, increasing overexploitation. Similarly, sustainable community management is hypothesized to become unsustainable as a result of efforts by state authorities or outside firms to enclose traditional collective property or impose new/foreign institutions. Related assertions posit that modernist development efforts to improve production systems of local people have led contradictorily to decreased sustainability of local practice and a linked decrease in the equity of resource distribution.

The environmental conflict thesis Increasing scarcities produced through resource enclosure or appropriation by state authorities, private firms, or social elites accelerate conflict between groups (gender, class, or ethnicity). Similarly, environmental problems become "socialized" when local groups (gender, class, or ethnicity) secure control of collective resources at the expense of others by leveraging management interventions by development authorities, state agents, or private firms. So too, existing and long-term conflicts within and between communities are "ecologized" by changes in conservation or resource development policy.

The conservation and control thesis Control of resources and landscapes has been wrested from local producers or producer groups (by class, gender, or ethnicity) through the implementation of efforts to preserve "sustainability," "community," or "nature." In the process, local systems of livelihood, production, and socio-political organization have been disabled by officials and global interests seeking to preserve the "environment." Related work in this area has further demonstrated that where local production practices have historically been productive and relatively benign, they have been characterized as unsustainable by state authorities or other players in the struggle to control resources.

The environmental identity and social movement thesis Changes in environmental management regimes and environmental conditions have created opportunities or imperatives for local groups to secure and represent themselves politically. Such movements often represent a new form of political action, since their ecological strands connect disparate groups, across class, ethnicity, and gender. In this way, local social/ environmental conditions and interactions have delimited, modified, and blunted otherwise apparently powerful global political and economic forces.

The target of explanation

Of course, each of these theses actually seeks to explain something somewhat different. While degradation and marginalization offers an explanation of why *environmental systems* change (because of accumulation), environmental identity and social movement research seeks to explain why *social systems* change (because of threats to livelihoods). This diversity of targets for explanation has been the source of some confusion in the field (Vayda and Walters 1999).

These differences reflect the historic development of the field. Research linking environmental change to political and economic marginalization emerged first in the 1970s and 1980s as an attempt to apply dependency theory to the environmental crises of the period (see chapters 4 and 7). The problematic effects of global and regional conservation efforts, including World Heritage Sites, national parks, and biodiversity zones, also became increasingly apparent in the 1990s, and political ecology on the topic benefited from a growing interest in the historical development of conservation (Chapter 8). Interest in environmental conflict soon followed, as many environmental issues became increasingly politicized in both regional contexts, from Love Canal to the Amazonian rainforest, as well as global ones, with the emergence of global agreements and debates on climate and biodiversity (Chapter 9). The last thesis, focusing on the new social activism that grew from all of the issues above, was placed squarely on the agenda by local people themselves, including Andean peasant movements, the Zapatistas, *chipko*, and a host of other movements (Chapter 10).

The diversity of political ecology research also results from innumerable, smaller, differing arguments addressing, among many issues:

- possibility for community collective action
- role of human labor in environmental metabolism
- nature of risk-taking and risk-aversion in human behavior
- diversity of environmental perceptions
- causes and effects of political corruption
- relationship between knowledge and power

Political ecologists, it would seem, don't agree on what they are trying to explain.

Certainly, however, these many topics and concerns overlap, and, as I hope to show by the end of the book, a coherent set of answers to these questions is beginning to achieve something of a consensus. Understanding how changing forms of knowledge, like computer mapping, lead to new systems of control over a forest, for example, will probably lead a researcher to ask: What are the concomitant changes in the behavior of foresters, and how does this create new patterns of actual forest ecology?

Moreover, in their linkages to local communities and non-governmental organizations, political ecologists, whether they are more interested in the biophysical or social aspects of a problem, have helped to build practical, detailed, integrated, empirical databases on all these diverse issues, recording land covers, farming practices, wildlife management systems, technological innovations and diffusions, local folk tales and oral histories, and informal markets and economies. These basic empirical findings help communities make decisions, aid in advocacy for social and environmental causes, and serve as a record to future scholars about the way things looked at the dawn of the twenty-first century.

The value of this last contribution, providing an historical record, is not a trivial one. Much of what we know about the political economy of the environment is bequeathed to us by political ecologists of previous generations. Indeed, political ecology can arguably said to be very old, since nineteenth- and twentieth-century environmental research in geography, anthropology, and allied natural and social sciences has a long critical tradition. Even before a semi-coherent body of political ecological theory emerged in the late twentieth century, many explicitly political practitioners emerged from the ranks of field ecologists, ethnographers, explorers, and other researchers. These represent the deep roots of the field.

2

A Tree with Deep Roots

Peter Alexeivich Kropotkin was born a Russian aristocrat in 1842, but by the time he died in 1921 he had become a globally known anarchist philosopher whose writings had done as much to explore the linkage between people and the environment as any in that tumultuous century. As an activist, a keen observer of nature, and a scientific explorer and ethnographer, Kropotkin was arguably the first political ecologist.

As a geographer, Kropotkin set out in 1865 to explore the most remote areas of the Russian Far East where the Sayan highlands border Manchuria. There were no charts of the region at this time and in preparing for the expedition he came across a map prepared by a Tungus hunter with the point of a knife on tree bark. "This little map," the explorer explained, "so struck me by its seeming truth to nature that I fully trusted to it" (Woodcock and Avakumovic 1990, p. 72). This trust for the environmental knowledge of local people was reinforced throughout his journey. Traveling for months with a local Yakut man, Kropotkin traversed 800 miles of rugged mountains. During his journey he encountered and described a wide range of farmers, herders, and hunters who all organized their lives to thrive under what urban Russians would have considered unthinkably adverse conditions.

Box 2.1 Denaturalizing politics and economy in Kropotkin's *Mutual Aid*

Darwin's earth-shattering thesis began to take on a life of its own not long after it was first presented. The notion that natural selection meant a struggle for existence between competing life forms was quickly extended to a more grim and entirely unfounded and unscientific claim that in social life the strong survive and the weak are naturally weeded out – social Darwinism. The hold this concept has on the public imagination remains strong (consider Herrnstein and Murray's *The Bell Curve*, 1994), despite lots of evidence that it is bunk (Mitchell 2000). It is also an idea that has been used to support opposition to progressive social services and assistance. So too, it opens the door to a range of false questions and alternatives: Should we overcome our "natural" tendency to struggle against one another through stronger central authority or higher ethical and moral sentiments? (Huxley 1896). Or worse: Why feed the poor when after all we'd all be better off as a species if they were dead?

For this reason Peter Kropotkin's century-old classic, *Mutual Aid: A Factor in Evolution*, is still timely and profound. Using his remarkable knowledge of animal ethology, coupled with his careful reading of available histories of social groups around the world, Kropotkin offers a solid anti-thesis to the social Darwinists' claims, rooted in solid empirical data; while competition may occur between species and between some individuals within species, cooperation between individuals is the key to survival and a central mechanism in natural selection.

While explicitly *not* arguing that the state of nature was "all harmonious," Kropotkin sought to draw attention to the way collective benefits accrue through coordinated efforts amongst ants, bees, white-tailed eagles, prairie wolves, marmots, and apes. This inherent sociability surely played a role in the "descent of man," therefore.

Of course the book reflects the evolutionary biases of the day. The ascent of civilization is assumed in the organization of the volume, with human societies historically stratified along an evolutionary ladder from "savages" (bushmen) to "barbarians" (Russian steppe village communes) to the "medieval city" (historic Novograd) to "ourselves" (Swiss alpine villages). Even so, at every step Kropotkin lays down vivid details of communal institutions and the successful coordination of activities in the absence of any kind of coercive authority. He does not argue that "savages" are any more or less cooperative than "ourselves." Rather, he demonstrates mutual aid in every form of social organization.

This position – somewhere between Darwin, Hobbes, and Rousseau – reflects not only Kropotkin's keen eye for the natural world, but also, of course, his explicit political project: to plot a way towards a technologically advanced civilization (Kropotkin 1985) based on equity (Kropotkin 1990), free from the hierarchic tyranny of the state (Kropotkin 1987), and in harmony with biophysical processes. By disallowing proponents of *laissez-faire* social systems to claim the natural high ground of evolution, he re-politicized the debate around natural selection, and in the process showed the danger of reductionist thinking. Kropotkin became one of the first real political ecologists.

Like his previous expeditions, this arduous trip reinforced his growing appreciation for "the constructive work of the unknown masses, which so seldom finds any mention in books, and the importance of that constructive work in the growth of forms of society" (Woodcock and Avakumovic 1990, pp. 59–60). The evidence amassed during these journeys, of plants, people, and animals making a living from the land, convinced Kropotkin, moreover, that the survival and evolution of species is propelled by collective mutual aid, cooperation, and organization between individuals.

The Determinist Context

Yet such research was by no means the norm. For the most part, early geography and nascent anthropology were the tools of social and political control, reproducing the political and ecological order that critical human–environment researchers would later challenge and undermine. Linking environment to society through a tradition of *environmental determinism*, scientific and field researchers were servants of colonialism and empire.

Rooted in the theories of nineteenth-century geographer Freiderich Ratzel and championed later in North America by the influential researchers William Morris Davis, Ellsworth Huntington, and Ellen Churchill Semple, the determinist approach maintained that geographic influences determined human capabilities and cultures, with its practitioners attempting to codify that thesis into scientific practice. Huntington was perhaps its most prolific exponent: "Today a certain peculiar type of climate prevails wherever civilization is high. In the past the same type seems to have prevailed wherever a great civilization arose. Therefore, such a climate seems to be a necessary condition of great progress" (Huntington 1915, p. 9). The empirical vacuousness of this thesis need not be belabored here. In even simple analysis, European and American environments have proved no more productive or inspiring for human life than any other (Blaut 1999, 2000). Where confronted with contradiction (e.g., "high" civilization in "bad" climate), Huntington and his colleagues generally retreated behind "complex" and "competing" factors and poorly defined trajectories of climatic change, while harsh climates were simultaneously used to explain the ingenuity of some groups and the cultural limits of others.

The implication of this theory in the perpetuation of global, imperial, racist rule by Euro-Americans should be immediately evident. By even asking the question: "Why are Anglos more productive, civilized, and advanced?" the fallacious assumptions have already been made that first, they are, and second, that it has to do with something inherent in the place or people involved, rather than being a consequence of historical and geographical interactions with the rest of the world. And in "scientifically" attempting to untangle the ancient question of heredity or environment, "Race or Place?" as Huntington put it in his classic volume *Civilization and Climate*, the fundamental political and historical questions of domination, colonization, and extermination are erased. In the answer, moreover, came a confident and scientific rationale for Euro-American dominance – *it's only natural*. Indeed, by rendering colonial domination an environmental inevitability, the practice of colonialism comes to appear *apolitical*.

This scientific thesis was quickly adopted in public service (Harrison 1999). Elementary geographic education during the turn of the twentieth century, in particular, was explicit on themes of environment and society linkage, with typically racist and colonial goals. A widely distributed text from the time, for example, *Guyot's Physical Geography* (1873), clearly asserts the association of continents and "ever-varying external conditions" with adaptive and functional purposes. "Each continent has, therefore, a well-defined individuality, which fits it for an especial function. The fullness of *nature's life* is typified by Africa . . . in the grand drama of *man's life* and development, Asia, Europe, and America play distinct parts, for which each seems to have been admirably prepared" (Guyot 1873, p. 121).

Each such region was further associated with clear and distinguishable "races of men." Of the "primary races" (white, Mongolic, and negro), the white race was held as "normal" and "typical," associated with the "refinement and culture of the European nations," and linked to the special function of Europe and America (Guyot 1873, p. 114). Whereas "the secondary races have contributed nothing to the present condition of mankind; and none of the existing branches have taken more than the first steps in civilization, except under the influence of the White or Mongolic races" (p. 118). This body of theory was crudely set in a roughly social Darwinist theory of selection, holding that geographic influences, acting through selection of superior specimens, created racial and cultural characteristics. This was accompanied by a great deal of pseudo-science, including most notably craniology and the comparative measurement of various body parts (Gould 1996). So too, a deterministic geography was supported by semi-theological theories suggesting the providential progress of divinely inspired dominion of the earth by Anglo-Americans (Livingstone 1994, p. 136).

The research implications for this kind of work were stultifying. By assuming the role of nature to be a determinant, fixed, and unidirectional influence, the complex influences of humanity upon non-human systems were lost altogether. As a result, nature was seen as a one-way force that determined cultural development, even at the very moment that the world of nature, ironically, was transforming under the processes of industrialization. Despite deleterious changes in air, land, and water resulting from economic and political developments of the era (smokestack industries, urban waste, and deforestation), nature was viewed during this period as beyond human influence.

A political ecological alternative

In this imperialist context, where environmental influences were understood to determine the superiority and inferiority of human races through competitive natural selection, and where human influences on the environment were viewed as unworthy of examination, counter-movements were growing. In the work of several early researchers a radical alternative to these dominant modes of explanation emerged, an incipient political ecology. Perhaps most prominent amongst these early dissenters was Kropotkin.

In contrast to the political and social conservatism associated with the geography of the period, Kropotkin's experience in the field brought him to renounce his princely title, to espouse a progressive policy of social cooperative anarchism, and to resist and dismantle the hierarchic social conditions of the time (Kropotkin 1990). In his work he sought simultaneously to tear down the socially loaded assumptions of contemporary, taken-for-granted, scientific knowledge, and to establish the empirical basis for an alternative model of social and natural organization. His research took a political ecological "hatchet" to the elitism and classism that pervaded natural science, while "seeding" the field with rich empirical investigations and normative visions of alternative futures.

Kropotkin's hatchet was aimed at the "scientific" socio-biological emphasis on competition that many other scholars saw in nature. Kropotkin argued that the case for competition as the central component of evolution was a product less of empirical observations of natural phenomena than of reading a social hierarchy into the natural world (Kropotkin 1888). He argued that there was nothing natural about competition, domination, or hierarchy; indeed the effort to naturalize them by seeing them everywhere in nature was to make unnecessary social hierarchies appear inevitable. In classic political ecological fashion, he deconstructed the dominant modes of science in his time.

Kropotkin searched throughout the animal kingdom and human history and pointed to cooperation being central to survival and selection, and therefore to evolution. His rigorous field observations made way for an argument not only about the state of nature, but also about the possibilities for society, free from domination, violence, and hierarchy.

Thus Kropotkin's approach to human–environment interaction sets a precedent for the kind of work that would follow more than a century later. The work resembles contemporary political ecology (and its progenitor, cultural ecology) in its focus on production, its archival and field-based empirical approach, its concern for marginalized and disenfranchised people, its interest in local environmental knowledge, and its concentration on the landscape as an object of explanation.

- A focus on *production* (farming, fishing, herding) as a key social-environmental process means taking seriously the notion that "the means of production being the collective work of humanity" (Kropotkin 1990, p. 14), the business of making a living therefore provides the most direct window into the mechanisms of social and environmental interaction.
- A rigorous *archival and field-based* empirical approach allows detailed observations of plant and animal life as well as historical social case histories from around the world (Kropotkin 1888).
- An explicit concern for *marginalized and disenfranchised* communities enables exploration of "institutions, habits, and customs" that, despite persistent exploitation by landlords and the state, locals prefer to maintain rather than adopt inadequate state-sponsored solutions "offered to them under the title of science, but [that] are no science at all" (Kropotkin 1888, pp. 260–61).
- A strong interest in the position and power of *traditional environmental knowledges* allows a pragmatic view of social and technological change. Though Kropotkin was a strong supporter of innovation, he insisted that the elements of progress could only be found in the existing resourcefulness of communities (Kropotkin 1985). The "hierarchical" forces of state and capital tend to crush "popular genius" (Kropotkin 1987).

- *Starting from the landscape* facilitates a grounded approach to social and political analysis, especially the influence of people on environmental systems. Ever the geographer, Kropotkin was as interested in environmental change as he was in social reform (Woodcock and Avakumovic 1990).

Taken together, Kropotkin's framework asserts that, left to their own devices, local subsistence production systems are generally cooperative and sustainable. It is only the overlaying of disruptive hierarchic forms of authority that lead to competitive and overextractive practices and a disregard for the environment. In Kropotkin's research, therefore, we see many of the elements of contemporary political ecology and the roots of cultural ecology as well.

This is not to say that this early work is without flaws. In particular, Kropotkin's distrust of the state – in any form – and his romantic assumption of popular cooperation are problematic. In an era when corporate power and global markets rival that of the nations, should progressive ecologists call for the state to be dismantled? Is cooperation the "natural" state of local production systems, or is conflict important in the history of social and environmental change? Even so, Kropotkin's call for a critical science of environmental sustainability and equity was compelling and direct, and forecast a synthesis of social and environmental research.

The Building Blocks

The elements of Kropotkin's critical social ecology were offered at a time when geography and a nascent anthropology were anything but progressive or emancipatory tools of social and environmental change. Even so, other contemporary researchers took critical positions both against those arguments for the "natural" character of an unjust social world as well as those that ignored the human influence on the environment.

Critical approaches in early human–environment research

The late nineteenth and early twentieth century produced a range of critical environmental approaches, at varying scales of abstraction, which sought to describe and analyze the patterns of human interaction with the environment. While few of them took the directly progressive and emancipatory approach of Peter Kropotkin, who boldly suggested that our ideas of nature are formed by our social condition, many of them wrestled with similar questions.

Continental critique: Humboldt, Reclus, Wallace, and Sommerville Perhaps earliest in this area was Alexander Von Humboldt, arguably the grandfather of modern geography, who is best-known for his empirical investigations of the physical world, which took him around the world during the early 1800s. Humboldt's travels brought him into contact with people making a living under a wide range of conditions and coping with varying degrees of political and economic hardship. These gave him an apparent appreciation for the political and economic context in which people make a living and cope in their daily lives.

His interaction with local producers also gave him a feeling for the unity of humanity and a distaste for the racist myth of natural difference. Though his five-volume *Cosmos*, which aspired to be a truly comprehensive physical guide to the universe, had only a scant few pages on humanity, and these dedicated to race, Humboldt was careful to insist that "while we maintain the unity of human species, we at the same time repel the depressing assumption of superior and inferior races of men" (Humboldt 1858, vol. 1, p. 358). Though sometimes invoking the racial language of the period and though clearly implicated in colonial-era exploration, Humboldt was from his earliest writings insistent that the "inequality of fortunes" between white colonials and indigenous communities could only be solved through equal access to both civil employment and fertile land (Humboldt 1811). These conclusions arose especially from Humboldt's experiences in South America, as did his sensitivity to traditional resource use practices and the implications of colonial economic systems for social and environmental reproduction.

In a typical example, Humboldt described at length the perilous decline of the pearl fisheries in the Cumana region of Venezuela, a unique resource whose fruits had been traded throughout the continent for generations. While allowing the possibility that tectonic forces (earthquakes and submarine currents) played some role, he was explicit that recent overfishing during the colonial period was probably to blame, since mercantile practice increasingly involved large-scale mining of the beds, so that oyster "propagation had been impeded by the imprudent destruction of the shells by thousands." The pearl-bearing oyster, he added, lives only nine or ten years, producing pearls only after the fourth year, making the mass extraction of the oyster (a boat might collect 10,000 oysters a week) extremely destructive and only marginally profitable. He further insisted that traditional native practice, opening promising shells one by one, sustainably supported a high-demand economy for the commodity before European contact (Humboldt 1852, pp. 191–4). Humboldt held the political history of the region to account for contemporary levels of destitution, underdevelopment, and environmental decline, rather than native practice or racial characteristics.

Like Humboldt, the French geographer Elisee Reclus was dedicated to comprehensive accounts of human and physical geography. His *The Earth: A Descriptive History* was only slightly less ambitious than Humboldt's *Cosmos* in its universal scope (Reclus 1871). The critical politics in his orientation towards human–environment questions were considerably more explicit, however. Like Kropotkin, he insisted that observation of human interaction with nature held the key to understanding society and insisted that "the sight of nature and the works of man, and practical life, these form the college in which the true education of contemporary society is obtained" (Reclus 1890, p. 10). He asserted, moreover, that eruptive political action against current systems of inequity – revolution – is part of evolutionary change in social/environmental systems. Combining an urge for justice, especially for workers, with a broader project of describing socio-ecological change, Reclus challenged the notion that contemporary social structure and ecological practice were the inevitable products of evolutionary selection.

These kinds of challenges to social domination and imperialism can also be seen grafted into the very roots of evolutionary theory. Alfred Russel Wallace, a British geographer and naturalist, simultaneously developed the theory of natural selection while elaborating a critique of social hierarchy and land management. Wallace's travels in Amazonia and the Malay Archipelago during the mid-1800s led him to investigate how geographic factors influenced the range of species, whether by enabling or limiting their distribution. The boundary he discovered, which passes through the South Pacific, separating the distribution of Asian animals from those of Australasia, still bears his name as "Wallace's Line" (Raby 2001). His experience also drove him to investigate how people indigenous to these regions made a living and classified the natural world. He would be remembered best, however, for his assertion that individual animals best adapted to their environments had the best chances for survival, thus influencing the emergence of differential adaptations. Several years of correspondence with Charles Darwin on the topic followed, after which Darwin's own *Origin of Species* (Darwin 1860) would be published. Thus Wallace became a co-developer of the thesis of natural selection, fundamental to evolutionary theory (Gould 1996; Raby 2001).

These more famous works, however, encompass only half of Wallace's concerns. Along with support for women's suffrage, workers' rights, and socialism more generally, Wallace's earlier experiences in land surveying led to an abiding concern for land planning and social reform of property rights. Having observed land ownership traditions in non-European contexts, Wallace became convinced that there was nothing socially or ecologically optimal about current tenancy arrangements in Britain and advocated thorough nationalization of land. With tremendous foresight, he anticipated public concerns for control of land to encourage historic preservation, development of parks, and limits on urban growth and sprawl (Clements 1983).

As noted earlier, these nineteenth-century political ecological critiques are all the more notable in light of the role that geographical and ethnological sciences were playing in the creation of empire. Humboldt critiqued racism and ecological degradation in the Americas in a way quite counter to the typical role of most geographers, who mapped and surveyed for military and civilian control (Capel 1994). More radical critiques like those of Reclus flew directly in the face of French geography, which advanced the notion of nationalist imperialism and viewed the expansion of empire, especially in Africa, as a cure for "decadent" and "insular" contemporary French society (Heffernan 1994). Though he held to a controversial spiritualism, Wallace linked evolution, social justice, and land management to offer a critical anti-racist alternative to emerging social Darwinism (Clements 1983). Together, these turn-of-the-century critiques prefigured contemporary political ecology by more than a hundred years.

A simultaneous European re-assessment of human impact on the land was also under way, but witnessed and articulated by an observer unusual during this period for both her gender and her background. Mary Fairfax Somerville was born in Jedburgh, Scotland, in 1780, and, gaining access to only the limited levels of formal education afforded to women in the period, became self-educated, making her own way through Ferguson's *Astronomy* and Isaac Newton's *Principia* (Patterson 1987). Authoring many scientific papers, her central contribution, *Physical Geography* (Sommerville 1848), was unusual for the time, owing to its emphatic insistence on the impact of humanity on land, rather than vice-versa. Though the book is marred with pejorative characterizations of non-Europeans somewhat typical of the time, it is also filled with strident critiques of slavery, of land theft from aboriginal peoples, and, most notably, of reckless degradation of environmental systems by people through overuse, extraction, and the introduction of alien species. In a remarkable counter-argument to Huntington's climatic determinism, Sommerville argued that humans, by altering watercourses, cropping, and forest clearing, had actually altered climates, anticipating such arguments in contemporary science by more than a century. At the same time her volume bemoans the reckless power of colonial states, which have driven indigenous people from their land and to the brink of extermination. Sommerville linked political and ecological destruction, urging reflection and caution. Nor was she unique in her contribution; large numbers of women naturalists in the nineteenth century set a similarly critical alternative tone for scientific exploration (Gates and Shteir 1997).

Such emergent political ecologies in Europe set the foundations for a century of work that is too large in scope to survey here. Francophone political ecology, whose continued rise coincided with the decline and fall of French imperial adventures in Africa and Asia, grew throughout the twentieth century from these solid critical roots (Whiteside 2002). Other contemporary European political ecologies, from the United Kingdom to Iberia, are also deeply rooted in the contributions of these early practitioners (Martinez-Alier 2002).

Critical environmental pragmatism As *fin de siècle* natural sciences in Europe were colliding with theories of society and fostering the emergence of critical sciences at the human–environment interface, a simultaneous movement in North America sought to break the hold of determinism. While the momentum towards a critical human–environment project had been halted with the "false start" it made in its embrace of determinism (Turner 2002), a contrasting school of human impact study was emerging, closely informed by traditions of pragmatism and utilitarianism. Led most prominently by George Perkins Marsh, a Vermont-born philologist with a lifetime of experience in the American diplomatic corps, a "new school of geographers" emerged in the late nineteenth century. For these researchers, the analytic challenge was to determine humanity's role in changing the face of the earth in order to preserve it for the future. Marsh wrote in a normative tone, insisting that responsible science and practical development required that conservation of the planet was essential, "thus fulfilling the command of religion and of practical wisdom, to use this world as not abusing it" (Marsh 1898, p. 7).

In his groundbreaking volume *Man and Nature* (later *The Earth as Modified by Human Action*) Marsh exhaustively listed the impact of human activity on degrading terrestrial ecosystems, rivers, lakes, and oceans, and traced the secondary impacts of such transformations on connected systems. In particular, Marsh was concerned with loss of forest cover, in terms of its effect on climate, erosion, and siltation of waterways. With remarkable foresight he anticipated the "invisible bonds" of ecology, pointing to the seriousness that declines in mayflies and aquatic larvae, for example, might have on the broader ecosystem (Marsh 1898, pp. 136–7).

With its concerns for human impact on the landscape and its focus on the effects of uncontrolled extraction on the reproduction and sustainability of complex ecosystems, Marsh's work was a precursor to political ecology. Even so, the work contains very little in the way of political economy or any focus on the way economic and political power is exercised to determine the rate and character of these problems.

His concern for the productive capacity of the ecosystem was not extended in any way to the local populations who had traditionally managed them. Indeed, in a revealing footnote, Marsh castigates the peasantry who "set fire to the woods and destroy them in order to get possession of the ground they cover" (p. 373 footnote), with absolutely no effort to place those actions in political or economic context. Why are peasants seeking to increase their holdings? What are the legal and institutional structures encouraging or dissuading such actions? What is the value of forest land?

There is also in Marsh's work a remarkable enthusiasm concerning the power and desirability of human "reclamation" of the earth. Writing on American forest plantations, for example, he fervently asserts that forest can and should be established anywhere and everywhere possible, specifically for its timber value. He further insists on the great social benefits of draining swampland (which today we call "wetlands") and straightening rivers, both practices that contemporary environmentalists abhor. His call for better stewardship of the environment was one that demanded more, not less, control over nature, especially by state authorities and private firms.

So while Marsh recognized the power of human economy to wreak environmental havoc, his faith in humanity's powers over nature led him to champion large-scale authorities and economies in a way that might have made Wallace, Reclus, and Kropotkin (not to mention late-twentieth-century environmentalists) uncomfortable. Even so, these themes – degradation, sustainability, and the culpability of human systems in the transformation of the earth – would all be central to political ecology a century later.

Clearly, an incipient form of critique lay in the works of many researchers working at the nature–society interface in the twilight of the nineteenth century. Political ecology would not appear, therefore, as if with a thunderous lightening stroke, full-grown, in the last decades of the twentieth century. But fully formed research into the politics and economics of environmental change, environmental knowledge, and the impact of state authority and the market on nature were yet on the horizon. Directed, empirical research into the geography and anthropology of the production and construction of nature would grow instead in other fields, specifically natural hazards and cultural ecology.

From sewer socialism to mitigating floods: hazards research

The politics and economics of environmental issues, though briefly eclipsed by environmental determinism in the early twentieth century, never fully left the scientific agenda. With an increasing recognition of the vulnerability of modern society to environmental perturbations like floods, earthquakes, droughts, and fires, as well as to the toxins of its own invention, a policy-oriented avenue of research began to open in the early twentieth century. Focusing both on "natural" and "technological" problems faced by human communities, hazards research took as its goal the rational management and amelioration of risk – defined as the calculable likelihood of problematic outcomes of human actions and decisions.

This approach emerged in the wake of increasing political urban activism around issues of environmental health and welfare, as settlement house workers and other activists, mostly women, rose to take stock of environmental conditions and the planning regimes that surrounded waste, water, and air. The Women's National Rivers and Harbors Congress, Women's Forestry Commissions, and Federation of Women's Clubs throughout America joined with socialist/progressive mayors, unions, and municipal leagues during this period to champion food safety and urban infrastructure reform. The research component of this "sewer socialism" was informal but thorough, and activist researchers like Alice Hamilton and Florence Kelly performed path-breaking street-level analyses of environmental hazards, revealing relationships between typhoid and plumbing, and between toxins and machinery and workplace injury and death (Bailes 1985; Darnovsky 1992). More formally and with greater analytic rigor, Jane Addams, along with her cadre of public university trained social workers, conducted the first-ever systematic assessment of the relationship between municipal garbage collection in Chicago's wards and local death rates (Addams 1910).

In formal academic circles this pragmatic approach to risk was later rendered in more apolitical terms. In a now classic example of the hazards approach, Gilbert White challenged the conventional way of thinking about and dealing with floods, calling for a rational and somewhat radical alternative. Writing his thesis in the early 1940s, White concluded that the traditional way of dealing with flood hazards – building more engineered structures – is expensive, irrational, and does little to deal with the underlying, fundamentally *human* problem. Better land use planning and changes in people's behavior could more easily mitigate future impacts of natural flood events (White 1945).

More than simply informing the practical question of flood insurance subsidies and dams, White had gained a valuable insight into human–environment interaction; the traditional distinction of those things *natural* from those things *social* is rendered particularly difficult when viewing the environment as a hazard. A flood is a hybrid human–environmental artifact, no more an act of nature than one of planning.

This powerful lever on the problem opened several decades of research into human adjustment to the environment, leading later researchers like Robert Kates and Ian Burton to venture the claim that the environment is actually becoming more hazardous as a result of human development, rather than less so (Burton et al. 1993). The implication that current economic and political structures increase the riskiness of natural events holds tremendous implications for how our society and personal lives are ordered.

But as the academic project of hazards research matured into the late twentieth century, it lost the critical momentum of its activist precursors and failed to form a robust and relevant theoretical account of social adjustment to the environment. By 1975 hazards researchers reviewing their own field concluded that future research strategies should elucidate the different problems and opportunities provoked by specific disasters – what are the distinctive properties of a tsunami, for example, and how best to prepare for one? (White and Haas 1975). Though new hazards (global warming, nuclear waste, ozone depletion) present new specific problems (Burton et al. 1993), this does not necessarily open onto clearer or more comprehensive understandings of environment–society interactions.

Yet the field left tremendous scope for critical insight and research. As White so profoundly discovered in the 1940s, improved planning by state agencies and individual farmers who lived in flood plains was retarded by the continued reinvestment in subsidies and massive investment in engineered structural solutions like dams and levees. But the urgent question raised by these results – why do structural solutions prevail in the face of better alternatives? – could not be answered within the hazards approach, which focused almost exclusively on individual choice, free markets, and rational regulation. Rather, the issue can only be addressed fully by examining the *political economy* of flood-plain investment, the role of capital in agricultural development, and the control of legislative processes through normative ideologies, vested interests, and campaign finance. Similarly, the risk of floods is not uniformly distributed through populations. Are poor and marginalized groups more vulnerable to such events? What is the role of power in the environmental system and its relationship to people?

These questions raise manifold research opportunities for critical scholarship. As Ben Wisner and Maureen Fordham ask on their radical hazards webpage *Radix* (http://www.anglia.ac.uk/geography/radix/index.html):

> How are populations made more vulnerable to these hazards by war, by government policies, by misguided development projects? What about the spiking incidence of domestic violence after hurricane Andrew in Florida and the Red River floods, both in the USA? What about the fact that 40 percent of all deaths from tornadoes in USA occur in mobile homes – inhabited by low income people?

But to engage these questions would require a fundamentally different view of social and natural process. In hazards research, humans are purposeful individuals who first perceive a hazard, recognize available alternatives, and then rationally adjust their behavior. If an individual behaves "irrationally," it is the result of cognitive biases, willful ignorance, faulty perception, or other personal and social-psychological "problems." The contextual forces that create unequal vulnerability and differential response, therefore, fall outside the concerns of traditional hazards research. As Michael Watts put it in his critique of hazards research and human ecology, "in spite of the recognition by Kates, White and others of the strategic importance of social causality, they have no social theory capable of addressing social processes, organization or change" (Watts 1983a, p. 240).

While such a theory was on the horizon, taking its contemporary form in "environmental justice," hazards research was largely unable to address many urgent questions in its absence. Even so, hazards research had placed urgent practical development problems into the purview of human ecologists, boldly suggesting that natural events were indeed quite social. Researchers were asking crucial questions, political ecological ones. A similar problem would emerge in the related field of cultural ecology.

The nature of society: cultural ecology

In contrast to the pragmatic policy approach of hazards, a separate group of modern scholars took as their focus an academic exploration of the development and expression of culture, especially on and within the environment. Cultural ecology, as the field would come to be known, approached human–environment issues ecosystemically: humans would be seen as part of a larger system, controlled and propelled by universal forces, energy, nutrient flows, calories, and the material struggle for subsistence. Unlike hazards, cultural ecology sought universal and generalizable rules of human–environment interaction. But like hazards, it would falter on the same conceptual and practical problem: accounting for and understanding change in a complex modern political economy. The crisis of explanation confronted by cultural ecology would become the fulcrum on which political ecology would be levered into prominence.

Historicism, landscape, and culture: Carl Sauer Interest in the historical development of cultures and human impacts on the landscape, it must be remembered, was not much cultivated in turn-of-the-century geography. This was so much the case, that the publication of an essay in 1925, which simply defined the objective of geography as the interpretation of landscape and humanity's role in changing that landscape, was considered a breakthrough.

That essay, "The Morphology of Landscape" by Carl Ortwin Sauer (1965a), led the field of geography into a new tradition of cultural landscape studies for several decades. Centered at the University of California at Berkeley, and shepherded by Sauer himself, this new school of scholarship directed itself to research into human use of nature, especially the impact of human activities. Historical and archaeological data were joined with geomorphologic and soil studies to create bold, long-term accounts of how places came to look the way they did (Speth 1981). Inverting determinism, historical landscape studies sought to explain the physical patterns on the land (forest cover, soil erosion, stream flows) in terms of human culture rather than the other way around.

Duel interests motivated Sauer's work in this regard: culture history and material landscape. The notion of culture history grew from an "anthropogeography" that was simultaneously emerging in the young field of anthropology. Under the guidance of Franz Boas, a German geographer turned ethnologist, this approach to culture insisted on a historicist approach to explaining how human cultures took the form they did. Rather than focusing on the functional-causal explanations typical of previous determinism, this approach focused on the emergence and adaptation of culture over time, diffusion of cultural traits, and interaction between cultures (Speth 1978). Sauerian human–environment research would concern itself with detailed study of the *how* of local cultures, less than the *why*.

Secondly, Sauer's concerns were directed towards what he viewed as the ecological crisis of Western civilization. His strongly normative view of human impact on nature, inspired at least in part by the work of Marsh, was explicit in its castigation of environmental degradation and its characterization of the modern commercial economy as unsustainable.

> To this review of some of the suicidal qualities of our current commercial economy, the retort may be that these are problems of the physical rather than social scientist. But the causative element is economic; only the pathologic processes released or involved are physical. The interaction of the physical and social processes illustrates that the social scientist cannot restrict himself to social data alone. (Sauer 1965b, p. 152)

His interest in these topics was probably formed during his work for the Soil Conservation Service (SCS) and the Michigan Land Economic Survey (MLES) prior to his arrival at Berkeley. These services probably helped to foster both his concern for the condition of the environment and his interest in the everyday affairs of working people (Leighly 1965). More than this, the work for the SCS and MLES set the stage for Sauer's dedication to the role of the social scientist in the "real world."

The economic urgency in Sauer's worldview is seldom reflected in his research, however. Very little attention in any of Sauer's contributions to cultural research dwells explicitly on the modern economy and its relationship to the environment. On the contrary, over his long career, Sauer is most commonly associated with the study of "archaic" or pre-modern cultural and economic contexts: pioneers in Illinois, early human occupation of the Americas, and settlement of Kentucky "barrens," as prominent examples (Leighly 1965).

Most significantly, however, Sauer established in the Berkeley school of geography a tradition of *fieldwork*. This empirical tradition sent researchers into the countryside and around the world, exploring the social world of people as expressed in their use of nature. This set a research agenda that would live on into contemporary political ecology.

Julian Steward: a positivist alternative The postwar period brought with it a change in the structure and direction of social scientific research. While the historicism of anthropogeography continued to be championed by researchers like Sauer, a new generation of cultural researchers became increasingly interested in an explicit, predictive *science*, one that sought laws with universal applicability using rigorous, quantitative investigation of cause and effect. Inspired by positivist approaches to social science – those that sought to establish scientific theories and laws of social behavior and function – these researchers wanted to move beyond descriptive histories and landscape studies. Without returning to crude determinism, anthropologists and geographers were searching for a science of culture.

In Julian Steward, they found an intriguing alternative. In contrast to Sauer and other historicists, Steward was driven by an interest in cross-cultural comparison. While the historicist notion of history creating culture, he claimed, avoids the problems of determinism, he insisted that it could not *explain*, and so must be rejected, in the hope that general processes could be discovered, and patterns that explained common, global, culture types might be sought. "The cultural-historical approach is . . . also one of relativism. Since cultural differences are not directly attributable to environmental differences and most certainly not to organic or racial differences, they are merely said to represent divergences in cultural history, to reflect tendencies of societies to develop in unlike ways. Such tendencies are not explained" (Steward 1972, p. 35).

Unlike the determinists, however, Steward was emphatic in insisting that environmental factors do not determine humanity. Instead, *human interaction with nature through subsistence and work* is the determinant and directing influence of environment on the social and cultural order. The theoretical basis of this form of human ecology was elaborated as a model of determinant factors, premised on the imperatives of making a living, extending outwards to more contingent and globally varying cultural features (Figure 2.1). At the center of human societies, Steward argued, is a "culture core," those fundamental features of human life, especially technology, necessitated by the conditions under which subsistence is achieved through farming, herding, collecting, and working more generally.

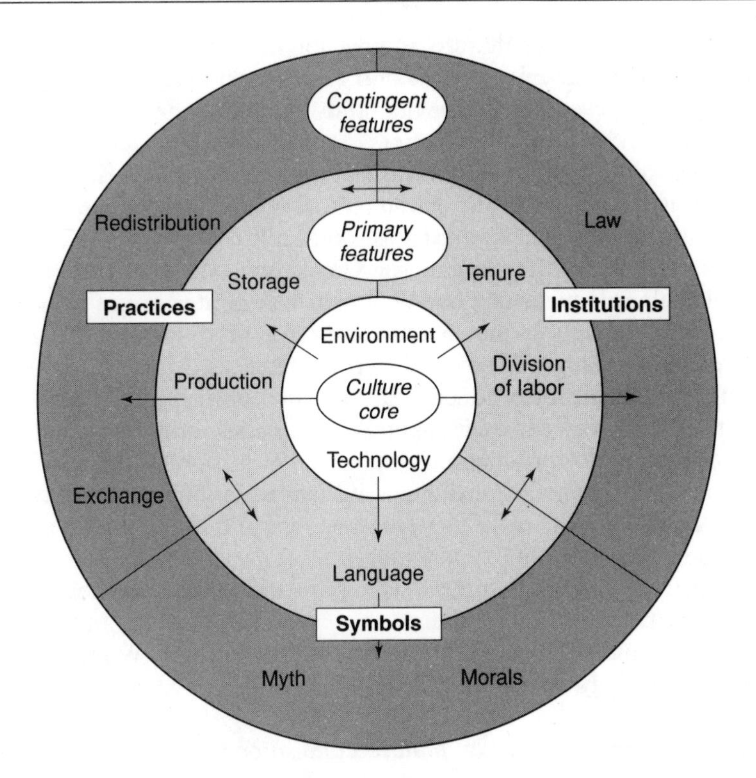

Figure 2.1 Julian Steward's cultural ecology

> The core includes such social, political, and religious patterns as are empirically determined to be closely connected with these arrangements. Innumerable other features may have great potential variability because they are less strongly tied to the core. These latter or secondary features, are determined to a greater extent by purely cultural-historical factors – by random innovations or diffusion – and they give the appearance of outward distinctiveness to cultures with similar cores. Cultural ecology pays primary attention to those features which empirical analysis shows to be most closely involved in the utilization of environment in culturally prescribed ways. (Steward 1972, p. 37)

These features were necessarily determinant of other, more contingent cultural factors. Thus semi-arid ecosystems, for example, do not *determine* the structure of a society in any simple way. But since all societies who hunt and gather in semi-arid ecosystems face similar production challenges, common social structural solutions might be hypothesized, influencing property relations, marriage patterns, food sharing, and other facets that together make up human cultural life. These patterns might give rise to a range of higher-order cultural functions, hierarchy, cosmology, and the broader morals and ideals of the larger culture group. Other cultural features may prove to be unrelated to subsistence practice and to have arisen through interaction with other groups or through "random innovations."

Methodologically, Steward's model formed a clear mission for researchers. By observing many cases of the same kinds of production and comparatively assessing those cultural characteristics that do and do not vary, researchers can scientifically understand what the ordered functions of social and cultural process might be. This regime further compelled researchers to study more "simply structured" and less complex social groups – hunter-gatherers, agriculturists, and herders. Though the same principles applied to modern industrial societies, it was reasoned, a scientific study should begin at the simple and extend only slowly to more complex cases (Steward 1972). Like Sauer, therefore, Steward's cultural ecology would focus on subsistence producers in rural areas, often in underdeveloped contexts. The implications of this for later political ecology cannot be overstated since contemporary work continues to remain overwhelmingly in the area of subsistence production in the third world.

Moreover, while Steward never directly references Marx or other Marxist theorists, in his concerns for production at the center of the "culture core," he comes quite close to articulating a materialist approach to culture, one fundamental to Marxian notions of the "mode of production" (Murphy 1981). For Marxian social science, the way in which people make a living and its determinant impact on forms of social organization and the flow of value allow explanations of history through careful examination of labor. This too would hold implications for research many years later, when the transition from positivist cultural ecology to the critical challenges of political ecology did not compel a reorientation away from production. Indeed, Steward's most prominent students, including Sidney Mintz and Eric Wolf, pursued more formal political ecology throughout their careers.

System, function, and human life: mature cultural ecology But long before such challenges emerged, the science of cultural ecology would develop into a complex and diverse field of study in its own right, propelled by simultaneous enthusiasm in geography and the emerging sub-discipline of ecological anthropology, and drawing upon an increasingly sophisticated body of concepts and quantitative methods (Grossman 1977; Hardesty 1977). In particular, the science of ecology provided the central analytical tools with which cultural ecologists would experiment. These were attractive because they allowed researchers to discuss human behaviors and practices in terms of their function and role in regulating energy and nutrient flows in a larger *system*, towards or away from homeostatic equilibria (Foote and Greer-Wooten 1968). In other words, by viewing humans as essentially the same as other plant and animal species, basic functional hypotheses could be proposed to explain complex cultural patterns. Rituals, kinship patterns, and traditional institutions could be evaluated to determine if they served an ecosystem function – one that helped regulate the general system of human–environment relations, maintaining stability.

In a classic example, Roy Rappaport, an anthropologist with years of field experience amongst the Maring people of New Guinea, proposed that major features of the Maring's complex culture could be explained by virtue of their role in maintaining ecosystemic balances. Specifically he hypothesized that periodic ritual warfare and pig sacrifice were the indirect response mechanisms to cycles of population growth and decline amongst both pigs and people (Rappaport 1967, 1968).

Other researchers used ecological concepts like "ecological niche" and "adaptation" to explore within and across group relationships (Hardesty 1975). Frederick Barth, working amongst diverse subsistence communities in the mountainous regions of Swat, Pakistan, suggested, for example, that inter-group relations, whether in terms of trade, indifference, or conflict, were regulated by the different niches that each group filled in the ecosystem (Barth 1956). Similarly, in the adaptation research of John Bennett, the practices of a range of ethnic groups on the northern plains of Alberta, including Hutterite farmers, Anglo ranchers, and Native Americans, were explained in terms of the different ways in which they made a living from the diverse resources of the prairie (Bennett 1969).

During the 1960s and 1970s such hypotheses were tested using increasingly formalized methodologies. Following on Steward's interest in a comparative science, cultural ecologists became increasingly interested in developing common metrics with which meaningful cross-cultural measurements and comparisons might be made. How can the patterns of resource use in the forests of New Guinea, for example, be meaningfully compared to practices in rural England or the Soviet Union?

The answer, researchers concluded, lay in the universal measures used in the study of terrestrial plant and animal ecologies: energy, nutrients, and biomass. The first of these, energy as measured in joules or kilajoules, was deemed most attractive; it flowed through all systems and could be used as a measure not only of productivity but also of efficiency (Rappaport 1975). This concern for efficiency would mark many scientific fields, propelled by the rise of systems theory in the last half of the twentieth century. By assuming that systems tend towards homeostatic balance, or that they shift between dynamic equilibria, the pattern of human action can be seen in a broader, systemic, and predictive order.

By following these flows of energy and matter, research into energetics could explore some profound and interesting questions. Has agricultural intensification through the green revolution in places like India and Egypt significantly altered the efficiency of production? Which is more efficient, Soviet collectivism or modern English smallholding? The results of this form of energetics research often cast traditional and "primitive" practices in a startlingly positive light. Bayliss-Smith's exhaustive and meticulous quantification of energy flows in agricultural systems around the world concluded that modern farming was remarkably inefficient, revealing the hidden ecological costs of fossil-fuel dependence (Bayliss-Smith 1982).

The adaptive practices of swidden (shifting cultivation or slash and burn) farmers, in particular, though long maligned by colonial officers and later development officials, were subject to careful scrutiny by cultural ecologists, who usually reached the conclusion that such farming systems were streamlined, effective, efficient, and environmentally benign (Conklin 1954; Geertz 1963; Dove 1983). Far from primitive and isolated, moreover, swidden was demonstrated to be well-integrated into complex market systems (Pelzer 1978), with recent work underlining its importance as a supplement for the poorest and most impoverished households (Hecht et al. 1988). These findings, presented in a world where modern, high-input, "green revolutionary" systems were being proposed as superior to those of traditional communities, sounded an important note of caution.

Finally, the cultural ecological approach is most notable for its serious attention to the logic of local people taken on its own terms, particularly their ecological knowledge and the relationships between that knowledge and environmental practice and the production of landscapes. Beyond the systems and adaptation approaches (most of which have lost prominence in the field over the years), research into people's logics and landscapes continues to thrive. The research of Robert Netting is most

Box 2.2 Netting's *Smallholders, Householders*: big things in small places

Robert McC. Netting's *Smallholders, Householders: Farm Families and the Ecology of Intensive, Sustainable Agriculture* is a wonderful contradiction, typical of one of cultural ecology's most enigmatic observers. Amassing a huge body of evidence and summarizing a lifetime of work, Netting's opus is concentrated on making only a single (but nevertheless important) point; intensive, small, peasant landholdings are inevitable, persistent, and sustainable.

The book is a blizzard of impressive detail from Nigeria and Japan to the Swiss Alps (long Netting's stomping ground). These particulars of farm strategy are harnessed to answer some big questions: Can the world's peasantry compete on regional and global markets? How much land is required for a farm family to survive? What systems of tenure allow small producers to thrive?

Like the bulk of Netting's work, the book debunks the high priests of intensification who tout that large farms with more machinery, operated by full-time agribusinesses, produce more and cheaper food. Where Netting discusses the question of a farm's size versus its productivity, for example, a hotly debated question with implications for the decollectivization of post-socialist farms and the agglomeration of corporate farmlands, he notes that "as is so often the case, cultural values are said to be responsible for economic inefficiency," keeping "traditional" and "spiritual" people close to the land despite the hopelessness of production arrangements. Nothing could be further from the truth, he demonstrates. The economies of scale enjoyed by some plantation industries and the marginal conditions of some dry areas notwithstanding, small farms show extremely high yields with relatively few capital inputs.

Netting's politics are also fairly clear. He finds leftist accounts of emergent inequality and the disappearance of "the little guy" just as unsupportable as the celebration of consolidation by big farm optimists. Intensive small farms, Netting insists, thrive in feudalism, capitalism, and late capitalism. And while "within-group" stratification of smallholders is a timeless reality, Netting insists that social mobility and opportunity in such groups is persistent – these high-population, high-intensity landscapes show remarkable social equality. These claims are contextual and debatable, of course, and do distract from troubling problems that smallholders will continue to face: unequal terms of trade, protectionism, and politically networked agribusiness. Even so, a material reading of the daily lives of smallholders shows a complex picture, one that defies grand theories about "the poor."

In this way, *Smallholders, Householders* does what many postcolonial critics have long urged: it abandons the creation of ethnographic accounts of the "other." Explicitly eschewing ethnology, Netting lays out his project from the start: "what follows is not an attempt to interpret 'culture,' a project of eliciting and perhaps creating meaning so grand that only the artist or the literary critic would confidently attempt it. Rather it examines a limited set of social and economic factors that are regularly associated with a definable type of productive activity" (p. 2).

Netting died in 1995. I regret that I never met or spoke with him. But this self-effacing passage seems to admirably capture his eminently practical voice.

notable in this regard. His lifetime of work with smallholders – "rural cultivators practicing intensive, permanent, diversified, agriculture on relatively small farms in areas of dense population" – demonstrated that though local production is immensely complex and highly variable, it operated in a sensible, rational, and relatively comprehensible manner. Whether examining the patchwork landscapes of field, pasture, and garden produced by Swiss peasants or untangling the complex property rights of the Koyfar of Nigeria, Netting consistently demonstrated that people close to the land acted with sophisticated ecological motivation and understanding to produce the world around them (Netting 1981, 1986, 1993).

Cultural ecology, in this way, opens the door to a range of productive questions, allowing a continuing exploration of the complex and sophisticated adaptations of people who had historically been characterized as backward. Related work continues to show the immense adaptive capacity of people, from the indigenous people of pre-Columbian North and South America to the producers of agro-food systems in the present day (Turner and Brush 1987; Turner 1990; Butzer 1992; Doolittle 2000; Denevan 2001). Researchers also continue to explore traditional ecological knowledge and management, with increasing recognition of the role of regional economic cycles in setting the terms of subsistence (Barham and Coomes 1996; Berkes 1999). By uniting highly specialized skills in agronomy, pedology, and hydrology with social and cultural exploration, cultural ecology has, moreover, created a model for integrative multi-disciplinary research in anthropology and geography (Butzer 1989; Turner 1989).

The incipient critical politics of cultural ecology is also readily apparent. Farming, herding, and hunting groups around the world, who have been characterized as primitive, conservative, and inefficient, become the focus of sustained and focused study, revealing the veracity and sustainability of their ways of life. It is the modern development state, by implication, with its high-input agricultural systems, its market orientation, and its urge to separate producers from resources, that appears primitive and inefficient. In the evolution of their work, cultural ecologists almost invariably, though perhaps not intentionally, have come to champion the most marginal and powerless groups, revealing the problems and limits of state and commercial power.

Even so, cultural ecology has been the subject of many criticisms over the years, both in terms of its concepts and its practices. Firstly, the excesses of the logic of adaptation, so central to cultural ecology, often lead to problematic reductionist conclusions, suffering from a fundamental teleological flaw: if people do it, it must be adaptive (Trimbur and Watts 1976). Indeed, the adaptation approach is focused specifically on assuming and demonstrating the ecological functionality of the most unusual cultural practices. The crude theories that developed from this approach propelled some truly bizarre and excessive claims. Aztec human sacrifice traditions, for example, the immensely complex socio-religious institutions of Mexico in the pre-Colombian period, were explained to be an adaptation to protein deficiencies for which human flesh was a crucial supplement. Leaving aside the fact that human protein demands could easily have been met with the maize–legume combinations of regional domesticates, the bold reductions necessary for such a claim were found to be unsatisfying and unrigorous even by supporters of the approach (Winkelman 1998).

This "neofunctionalism" was further criticized for its crude use of the concept of "carrying capacity," which uncritically assumed that there are given limits to human population density, despite extensive and growing evidence to the contrary (Behnke and Scoones 1993). The assumptions, moreover, that a given subsistence population could be analytically bounded also posed difficulties, as did the short time-scales of research over which arguments for long-term adaptation were made (Orlove 1980).

So too, neofunctional cultural materialism, as championed by anthropologists like Marvin Harris, has been overturned, often simply through rigorous field research. Harris, for example, argued that the cow became sacred in India because of the ecological value of its protein provision and agricultural traction power (Harris 1966). Highly inconsistent data and questions of cause and effect in cattle protection undermine any such simple explanation (Simoons 1979; Freed and Freed 1981). Do adaptive uses lead to taboos creating surpluses or does the surplus of animals lead to adaptive uses? As adaptation researcher Alexander Alland (1975) once insisted, the worst cultural ecology in this way represents little more than "just so stories" (p. 69).

Another ongoing criticism of cultural ecology centers on the degree to which it has been, and remains, parochial in its outlook, focusing almost exclusively on underdeveloped rural contexts. In an urbanizing and interconnected world, such a focus seems out of step with contemporary concerns and the globalizing realities faced by the same local producers that cultural ecologists claim to understand. By largely ignoring first world contexts and urban localities, cultural ecology is arguably increasingly less relevant.

Most problematic, the thrust of some cultural ecological argument explicitly naturalizes and, by implication, legitimizes what can be seen as contingent social behaviors and practices, recalling the socially and politically disturbing features of determinism. If the Native Americans of Bennet's *Northern Plainsmen* fill an "adaptive niche" by living at the edge of subsistence, scavenging at the periphery of the larger economic and ecological system, the implication is that such a status is natural, and not the result of land seizure, political marginalization, discrimination, and decades of exploitation (Bennett 1969).

The politics that both make up and constrain the daily life of such people, who are perpetually engaged in social and ecological conflicts over subsistence, are little in evidence in this work. This disinterest in resource politics, in the end, often makes it difficult for cultural ecologists to explain the outcomes they observe in the world. Even where truly visionary cultural ecology has called attention to looming development-driven crises, as cultural ecologist William Denevan did with remarkable insight for the Amazon as early as 1973, the limits of the approach, like that of hazards, are established by the absence of theoretical tools to address the larger political and economic climate.

Beyond land and water: the boundaries of cultural ecology

These limits are perhaps no more clearly seen than in *Between Land and Water*, Bernard Nietschmann's groundbreaking study of social and ecological change along the Miskito coast of Nicaragua. Nietschmann was a naturalist and by all accounts a lover of sand and sea, but with a strong interest in the workings of culture and a commitment to the scientific study of development problems. In 1968 he departed for a small community of Miskito Indians in the village of Tasbapauni on the Pearl Lagoon on the Caribbean coast of Nicaragua. Equipped with all of the robust tools and theories of cultural ecology, having been trained at the University of Wisconsin by William Denevan, a senior researcher in the field, Nietschmann intended to study subsistence strategy and change along the coast. The study would be extensively quantitative and would involve careful measurement of crops and game, with an eye towards exploring energy inputs and outputs, especially in terms of the harvest of green turtles, which were crucial components of Miskito subsistence and livelihood: classic cultural ecology.

Nietschmann's extensive and detailed quantitative conclusions are complex. His work concluded that the Miskito depend on hunting and fishing as key supplements to crop subsistence, since they provide dependable food security and consistently productive yields. Specialization by individuals in hunting or fishing, he maintained, was in part a result of expensive equipment costs and because mastery of the complex knowledge required was difficult (Nietschmann 1972). Similarly, Nietschmann recorded the complex and sophisticated systems that governed sharing of meat catches, examined in terms of the cultural role of redistribution in the reproduction of the social order and availability of proteins (Nietschmann 1973). All of these bear the traditional marks of cultural ecology's questions and answers.

But things were not in homeostatic order, by any means. The monetization of the local economy had redirected flows of exchange and harvest of hunted wild animals. Specifically, the trading of sea turtle meat and other products, though a practice dating from at least the early seventeenth century, had radically accelerated in recent years. This brought with it a breakdown of social reciprocity, an acceleration of harvest, and decline of turtle resources. This decline fed a spiral of overexploitation and capitalization with serious social and environmental implications (Nietschmann 1973).

Ultimately, Nietschmann concluded, the fundamental problems of Miskito subsistence and the emerging livelihood crises along the coast were not related to the metabolism of the internal ecological system – the governing system imagined to be so important in systems approaches of cultural ecology – but the broader global market. As he explains in a compelling narrative account of his life in the field, *Caribbean Edge*:

> These green turtles, caught by the Miskito Indian turtlemen off the eastern coast of Nicaragua, are destined for distant markets. Their butchered bodies will pass through many hands, local and foreign, eventually ending up in tins, bottles and freezers far away. Their meat, leather, shell, oil, and calipee – a gelatinous substance that is the base of turtle soup – will be used to produce goods for more affluent parts of the world. (Nietschmann 1979, pp. 173–4)

Concerned not only about the lives and livelihoods of the Miskito, with whom he had developed a strong rapport, but also for the turtles themselves, Nietschmann began to ask new and pressing questions. How much more overextraction might be expected before the Miskito respond economically or politically to their position in Nicaragua's political economy? Did struggle lie ahead? Could Miskito systems of production function with one foot in subsistence and the other in the market?

Evidence on the relationship between social conflict, markets, and turtle population decline was clear, especially in terms of traditional systems of reciprocity:

> Tension is growing in the villages. Kinship relations are strained because of what some villagers interpret as stingy meat distribution. Rather than endure the trauma caused by having to ration turtle meat, many turtlemen prefer to sell all of their turtles out to the company and return to the village with money, which does not have to be shared. However, if a Miskito sells out to the company, he will probably be unable to acquire meat for himself in the village, regardless of kinship or purchasing power . . . the situation is bad and getting worse. Individuals too old or too sick to provide for themselves often receive little meat or money from relatives. Families without turtlemen are families with neither money nor access to meat. (Nietschmann 1979, p. 186)

Nor would the political imperatives and entanglements of Miskito livelihoods in Nicaragua end there. The Nicaraguan Sandanista government of the 1980s, during their protracted conflict with US-supported guerrillas, enacted price controls and land seizures throughout Miskito territory, which further highlighted the marginal position of the community both within the global economy as well as within their own national polity (Nietschmann 1989). Nietschmann was compelled to address these issues, and would do so both as a researcher and as an activist apprenticed to the Miskito community.

The limits of progressive contextualization Though this struggle would drive research for the rest of his life (he died in 2000), the tools of Nietschmann's science did not seem to fit the range of questions he faced. Even as he had walked through a political doorway, Nietschmann had hit a conceptual wall. Restricted to research tools like organicism, function, adaptation, and equilibrium, further understanding could not cross the barrier point where markets meet subsistence and where the same local populations carry out the creation and destruction of the environment.

Cultural ecology offered one more conceptual instrument for understanding such complexity. Andrew Vayda (1983), writing in the early 1980s, proposed that explanation of people–environment interactions follow a path of "progressive contextualization," where human–environment interactions are explained "by placing them within progressively wider or denser contexts" (p. 265). The predicament of the Miskito, and communities like them, can be best explained by describing the changes and conflicts in their production system, while slowly refocusing the analytical lens to understand the social context of decisions, the economic context of those social systems, and the political context of that economy. Nesting local interactions within larger and larger scales, Vayda argued, leads to an understanding of driving processes in an empirical and inductive way.

But to describe, in an *ad hoc* and ultimately atheoretical fashion, does little to *explain*, to answer the question *why*, and reveal underlying and persistent processes. Why are turtles declining? Because of overfishing. Why is overfishing occurring? Because of changing markets. But why are markets changing? And what is the overall relationship between markets, state authority, local power, and ecological cycles of production and decline? The interactions between state institutions, coercive social

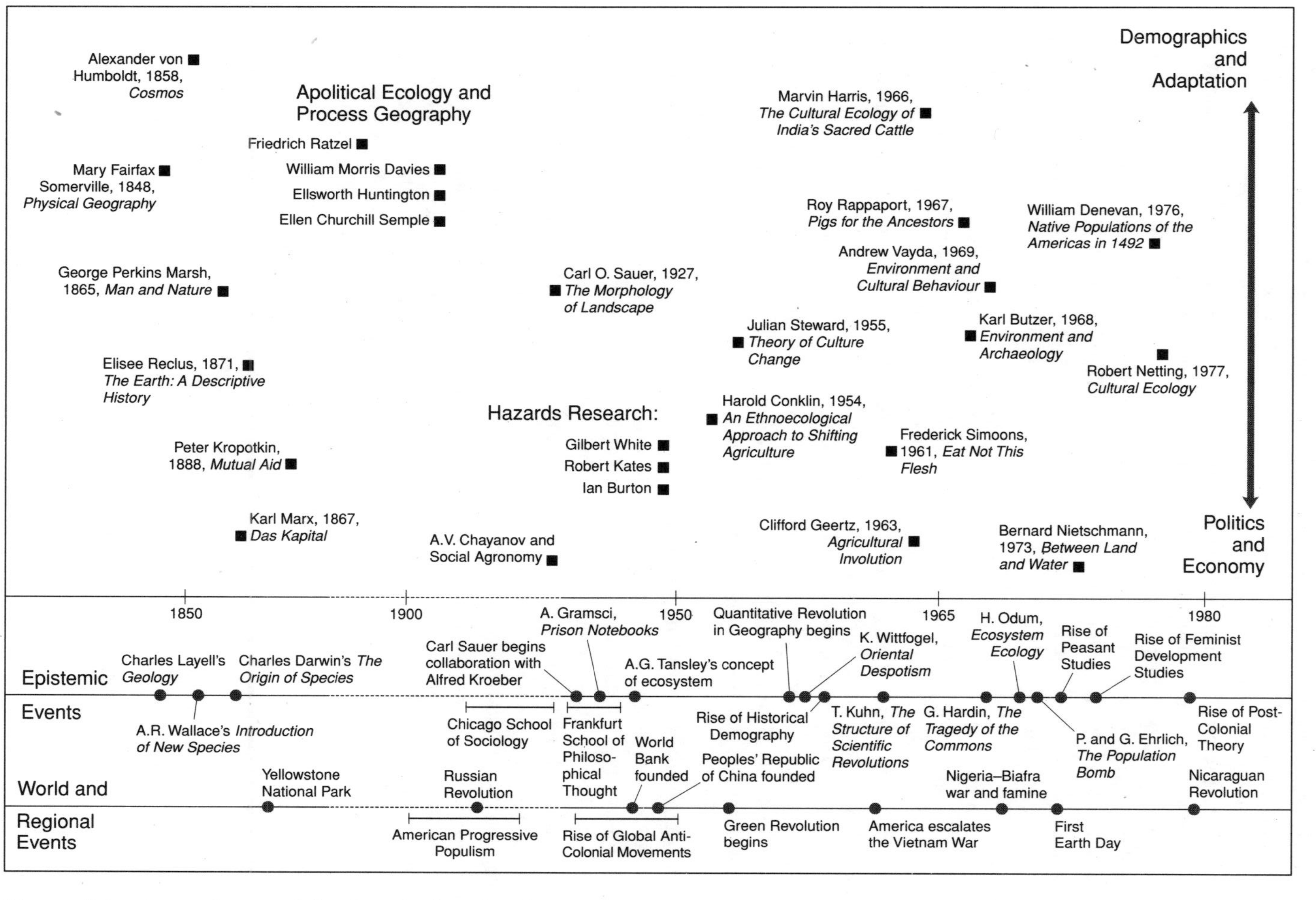

Figure 2.2 Antecedents and the rise of cultural ecology

relationships, commodity markets, subsistence, and natural resources were dynamics that required new theoretical tools and categories, not simply fuller description. This is especially true if the analyst wants not only to describe changing human–environment interactions, but to *change* them as well.

So too, the political role of the researcher in representing and interacting with the groups with whom they work had so far received little discussion in human–environment study. What are the obligations of the researcher to the researched? What are the inherent power relations that create problems in that relationship? Who can speak for whom? To whom is research speaking? To what end? These issues, though inherent in the work of hazards researchers and cultural ecologists, had received little serious attention. They would require far more elaboration before practical and progressive work might be done.

Like many researchers before him, Nietschmann was beginning to do what we now call political ecology. He had argued that Miskito articulation with global political economy had simultaneously created reconfigurations of social systems governing redistribution, cultural standards governing resource management, and environmental systems governing the populations of wild species. He had found change, but not the change he thought he would find. And as in Humboldt's observation of pearl fisheries, Sauer's anxiety over commercial economies, and White's examination of "irrational" flood policy, the theoretical tools to explain *why* such changes occur, which might help to steer both research and activism, were not yet fully formed.

Taking the plunge In sum, the general argument I am presenting here, insofar as the history of several fields can be used to draw any coherent lesson, is that critical politics in environment–society research are not at all a new thing. Indeed, from the very origins of evolutionary theory, through the complex social and ecological revolutions of the late nineteenth century, into the era of technocratic intensification and urbanization, researchers have continued to articulate a relatively coherent program of political ecological research. This work, from the anti-authoritarianism and anti-commercialism of Kropotkin and Sauer to the local rationalism of White and Netting, has consistently interrogated the logic of local production, the value of local knowledge, the environmental costs of regional and global change, and the power-laden impacts of socio-environmental change (Figure 2.2). As I have tried to show here, however, the consistent problem has been the absence of an integrated set of critical concepts, methods, and theories from which to explain problems and upon which to build alternatives.

Such critical tools, however, lie close at hand. And in the explosive political and ecological events of the 1970s and 1980s, these would find articulation in the increasingly formalized field of contemporary political ecology.

3

The Critical Tools

In 1978 the government of India, working with a large number of NGOs, set out to modernize the cattle breeding system of the poorest communities in Orissa, a state long plagued by drought and poverty and highly dependent on the milk proteins their few cattle provided. As reported by journalist P. Sainath, the project was greeted with tremendous enthusiasm by all those involved. The region's poorest households were given a free cow, impregnated with Jersey semen to create offspring with high milk yields. They were further provided with an acre of free land to grow trees for animal fodder. Finally, they were provided with minimum wage payments for their labor with the trees and animals until the new modernization project got off the ground and the benefits of purebred Western animals were experienced by all, in the form of plentiful milk, good sales, and a sustainable future (Sainath 1996).

By 1990 the project was in a shambles. Most villages across the region were now without any animals, including not only the new hybrid but also their traditional subsistence beasts for milk and burden-bearing. The region's traditional Khariar bull, long treasured and outbred for profit, was biologically extinct. The trees, once harvested, had died. The land had been reclaimed for marginal food crops. Modern development had made local people more destitute and had depleted much of the region's environmental resources and faunal biodiversity. A project intended to decrease outmigration from the area had increased it, forcing more independent producers into low-wage labor (Sainath 1996).

What had gone wrong? Clues to the failure of the system can be found by employing the principles of hazards and cultural ecology. Viewed in this light, the periodic hazard of drought in South Asia has resulted in traditional risk-reducing breeding systems that balance production and survival in livestock species. The traditional use of marginal fodder resources produced consistent and steady regional milk exports from the region, if not large ones. Land management was thus traditionally well integrated with climate variability. So too, stud bulls were commonly treated as community and village property to optimize access to key resources as well as production levels, while diversifying genetic stock. Plantation choices mixed food as well as feed crops, depending on community grazing lands for livestock inputs. The result was a system where milk production in Indian grasslands was roughly a remarkable 1.85 tons per hectare, compared to a global average of 0.137 (Crotty 1980; George 1990).

From this point of view, the project directors in Orissa did *everything* wrong. By introducing an American stud animal that was ill-suited to the climate, they all but assured its decline during periods of environmental stress. Insisting on purity of the genetic material, they slaughtered all local stud bulls so as not to dilute the breeding program. Producers who grew both food and fodder on their allotted lands had their vegetables torn from the ground (Sainath 1996). Despite historically communal breeding systems, animals were privately dispensed, as was fodder-producing land. The fodder tree species choice, *Leucaena leucocephala*, while a productive species, is not suited for all environments (Hocking 1993). So from a cultural ecological point of view, as well as from that of hazards adjustment, the project was poorly designed; indeed it was insane.

But the larger questions still loom. Why did the development authorities make the decisions they did? Why were local practices dismissed? Who sponsored foreign high-yielding animals, ones that required high levels of purchased inputs? Who benefited from this program in terms of animal and semen sales, international consultancy fees, and administrative salaries? How does the state's relationship with non-state actors and global markets direct the choices made? Does this represent a larger development trend in the articulation of local production systems with globally distributed and marketed trees and animals? How did independent landowners become landless workers? Who claimed their abandoned farms? These questions, hallmarks of political ecology, remain.

As with Nietschmann's Miskito, the tools of cultural ecology and hazards, though crucial for describing such ecological systems and problems systematically, are insufficient for asking and answering the pressing multi-scale questions of development-era environmental change. The emergence of a wide range of crucial theoretical concepts in recent decades – drawn from common property theory, green materialism, peasant studies, feminist development studies, discourse theory, critical environmental history, and postcolonial theory – constitute a new and robust toolkit to directly tackle these questions. They together form the eclectic equipment of political ecology.

Common Property Theory

One of the first and most essential contributions to a contemporary political ecology is common property theory, which rests on the understanding that fisheries, forests, rangeland, genes, and other resources, like many of the environmental systems over which struggles occur, are traditionally managed as collective or common property. Indian pastures, like Nietschmann's turtle fisheries, White's rivers, and Humboldt's pearl beds, are all complex ecological systems that are difficult to divide into individual units of ownership – to "exclude" in the language of economics – owing to their spatial and temporal variability. But where private benefits are accrued at a cost to the group, there is a potential to overgraze rangeland, pollute rivers, overextract fish, and otherwise use resources unsustainably. Clearly many of the environmental systems of interest to cultural ecologists and other environmental analysts seem to fall into this broad category of vulnerability. Moreover, since the possibilities for environmental degradation under these social and ecological circumstances are high, problems like declines in pearl beds or turtle populations might be explained as tragic outcomes of failures in collective management.

Local management structures, rooted in local knowledge of such environmental systems, however, commonly provide rules of use that maintain subsistence and renewal of these community resources. Community managed resources in fact thrive around the world. A widening international interest in the operation and function of those rule systems emerged in the 1970s and 1980s, an interest that was concurrent with the rise of contemporary political ecology.

This body of research grew out of a response to the conventional wisdom in the West, wisdom rooted in the premise that private gains might hold social or ecological costs, and which held that collective use of resources tended inherently towards abuse and degradation. Codified into a socio-economic theory – "The Tragedy of the Commons" – this conventional wisdom insisted that only centralized regulation or privatization could solve the dilemma of collective resources (Gordon 1954). "Picture a pasture, open to all," Garret Hardin begins in his classic statement on the question:

> It will be expected that each herdsman will try to keep as many cattle as possible on the commons. Such an arrangement may work reasonably satisfactorily for centuries because tribal war, poaching, and disease keep the numbers of both man and beast well below the carrying capacity of the land. Finally however comes the day of reckoning, that is the day when the long-desired goal of social stability becomes a reality. At this point, the inherent logic of the commons remorselessly generates tragedy. (Hardin 1968, p. 1244)

Robert Wade presented it with greater clarity. The choice facing community resource users is:

> either to cooperate with others in a rule of restrained access or to not cooperate. The argument is that each individual has a clear preference order of options: (i) everyone else abides by the rules while the individual enjoys unrestrained access (he "free rides" or "shirks"), (ii) everyone, including himself, follows the rule ("co-operates"); (iii) no one follows the rule; he follows the rule while no one else does (he is "suckered"). Given

> this order of preference, the stable group outcome is the third-ranked alternative, unrestricted access to all in the group. The second-ranked alternative, with mutual rule-bound restraint, is more desirable. But this is not stable equilibrium, because each individual has incentive to cheat and go for the first ranked alternative (restrained access to all but him). Even if it turns out that no one else follows the rule, his cheating at least ensures that he avoids his own worst alternative – following the rule while no one else does. (Wade 1987, pp. 97–8)

Under this logic, individuals, assumed to be seeking individual benefit, will invariably take as much as possible from collective resources. Since the costs of that extraction, in reduced returns due to overgrazing, overfishing, or overcutting, are shared between all members of the community whereas the benefits are accrued alone, the inherent logic is to continue and indeed to accelerate individual extraction. When enough individuals behave in that fashion, environmental destruction is inevitable. The only options are centralized coercion or privatization. In the first case, a state entity, exogenous to the group, forces stocking rates on the herders, fishers, or woodcutters. In the latter case, the commons is divided into pieces and distributed between individuals, so that overuse of the resource will be immediately felt by the perpetrator and can be individually rectified.

The argument for the tragedy of the commons is tidy, internally coherent, persuasive, and meritorious given its assumptions. And using rational choice theory and game theory – where logical individual actions are modeled in anticipation of the actions of others – various scenarios of this sort can be tested. Consistently they seem to produce the same result. Failure occurs where individuals seek personal benefit in environmental systems and costs are "externalized" to the group.

But empirical evidence compiled for the last three decades shows less support for such a model, and time and again evidence of collective stewardship appears in the management of resources ranging from fisheries from Maine to Turkey, pastures from Morocco to India, and forests from Madagascar to Japan. While "tragedy" theory suggested failure, the literature was filled with "exceptions," locally organized techniques, rules, and decision-making structures that organized extraction, defined user communities, and maintained harvests and yields. The empirical record on common property management is far too large to survey here, but the accumulated case material is impressive (see National Research Council 1986; Feeny et al. 1990; and Burger and Gochfeld 1998).

The search during the 1970s and 1980s for an alternative theory, therefore, became an imperative for international and comparative social science research. How to account for these successes? When do they work? What makes them fail if and when they do? The theory that emerged would challenge the basic assumptions of the "tragedy" thesis, first by suggesting that commons users are not isolated decision makers but in fact live in communities where they can mutually monitor and communicate, and second, that the tragedy "game players" can watch outcomes unfold and adapt their decisions in later "rounds of play." Following the pragmatic tradition of institutional economics, this alternative theorization suggests that in fact myriad solutions to the problem exist, if conditions allow for negotiation and iterative observation of outcomes (Commons 1990).

Success of collective management, theorists maintained, is a result of the fact that such commons are not unowned (legally, *res nullius*) but are in fact commonly held property (legally, *res communes*) (Ciriacy-Wantrup and Bishop 1975). Failure of collective management, by contrast, merely represents failures in the specific structure of rules that govern a collective property, by virtue of increasing scarcity or value of the resource or alterations in local social structure and culture. Recovery of sustainable management is a task of crafting new and better rules, not one of slicing up the commons into private bits, nor imposing strong-arm central authority (Ostrom 1990, 1992; Ostrom et al. 1993; Hanna et al. 1996).

Again for problems like those facing the Miskito, where broader economic forces were transforming the harvesting of traditional resources, this approach provides some useful lessons. Overfishing of mobile resources, like sea turtles, is by no means an inevitable or even a common outcome of collective ownership and management. Communities like the Miskito had sustainably harvested such resources for generations, through clearly defined systems of social sanction, redistribution, inclusion, and exclusion. Explanation of the failure of the sea turtle fishery lies, therefore, in the problem of how the rules work, and whether they can adapt to socio-economic change; the Miskito, as rational decision-makers, might yet craft new rules.

In this way, most responses to the "tragedy of the commons" took the question on its own terms, proving empirically that given the opportunity to negotiate and given the proper structure of rules, degradation was by no means the inevitable result of collectivity. Rational choice, therefore, was used to form an apparently apolitical theory of environmental commons.

Other critics were bolder, however. They held that the increasingly capitalized economies were radically altering the social and political circumstances of the players of these commons games. Indeed, as Muldavin phrases it, the entrance of coercive states and new markets into social economies like that of the Miskito results in the *appropriation* of communal capital away from locals and into the hands of elites, non-residents, and other distant parties (Muldavin 1996). The "tragedy of the commons," moreover, by placing the fault of degradation at the feet of disempowered local communities, actually disguises and supports this outcome. This observation would become fundamental to political ecology.

An *apolitical* theory of the commons, therefore, though attractive, is inadequate. Multiple scales of power and diverse players acting on local commons are unexamined and the multi-scale structure of the economy unacknowledged. The broader historical trajectory of socio-economic change is ignored. Moreover, by continuing to insist on the apolitical nature of the problem, such approaches to the common property problem reinforce the normative assumptions of rational choice "tragedy" approaches. Practical action is limited to internal "rule crafting," which does not challenge the more fundamental economic forces at work. A more ambitious and explicitly political thesis would be required, drawing on materialism and political economy.

Green Materialism

In the same year that Peter Kropotkin returned from Siberia to begin his work as an activist and philosopher, 1867, Karl Marx published the first volume of his three-volume masterwork *Capital*, cementing what would become a parallel but distinct line of environmental investigation emerging in philosophy and economics: Marxist materialism. For Marx and Engels, who observed the industrial revolution with both awe and concern, the degradation of the environment was a fundamental feature of capitalism. The politics of the environment was, therefore, linked to the politics of class struggle, industrialization, and capital accumulation.

Though Marxist philosophy and economics are complex and provide a range of tools, two key precepts in particular would later have a great influence on the development of political ecology. The first is the assertion that, according to Marx, social and cultural systems are based in historical (and changing) material conditions and relations – real stuff. Following from a long line of philosophers, including the Greek philosopher Epicurus (*c.*341 BCE) whose work was the topic of Marx's doctoral thesis, materialists challenge the notion of idealism, which holds that philosophy, consciousness, and ideas are the engines of history, constituting the world and its transformation. In contrast, the materialists argue that the way humans interact with the world of natural objects provides a "base" upon which law, politics, and society are founded and around which they are given form. As production and the relations of production (social relationships that govern how objects, food, and goods are made, harvested, and assembled) change, society changes as a result (Foster 2000). Such a notion echoes Steward's concept of the "culture core" described above.

The second key notion is that capitalist production (a specific and recent kind of production) requires the extraction of surpluses from labor and nature. As that extraction increases in intensity, contradictions emerge that provide barriers to further growth, bringing a possible end to capitalism. For materialists, environmental degradation is therefore inevitable in capitalism but also one of its fundamental weaknesses (O'Connor 1996).

Materialist history

This materialist view of history lends itself to investigations of the relationship between nature and society. If forms of social organization are rooted in production (how things are made), they are, by implication, ultimately explained by how people use nature. For many materialists, this has meant a broad, general, and all-encompassing theory of history, explaining how one society transforms into another. For these historians, the central concept in understanding such change is the "mode of production." Simply put, a mode of production is a combination of key social and material elements; these elements are constant, and include labor, technology, and capital, but their interrelationships, combination, and recombination are in constant flux, leading to differing ways of making a living from nature, and changing organization of society across history and over space. In a pre-capitalist mode of production, for a simple example, shoemakers own their own capital/equipment and make shoes, selling the finished product to buyers. In a capitalist mode of production, on the other hand, the machinery of shoe manufacture is owned by a capitalist; workers do not own that equipment, and have nothing but their labor to sell, being paid for time spent making shoes that belong to the owner.

The transformation from one system to another is driven by internal changes and "contradictions" in the system, leading to ongoing historical struggles that create new ways of organizing labor and nature (Althusser and Balibar 1970). Contradictions are elements in the systems whose relationships to one another are necessary but inherently at odds, eventually leading to crisis, fracture, and collapse. Materialists assert, for example, that because surplus value must be constantly extracted from workers and from the soil to landlords and commodity traders, the conditions (human health, soil quality, nutrients) required to maintain that production cannot be sustained, leading to a crisis, and possibly socio-economic change.

Where one mode of production encounters another – kin-ordered subsistence like the Nicaraguan Miskito encountering global turtle markets, for example – a process of "articulation" follows. The two systems struggle, forming a hybrid outcome in which producers may have one foot in global markets and another in subsistence production exchange systems (Hindess and Hirst 1975).

Many theorists in this area are quick to point out that there is no teleological end (inevitable trajectory) for all of this social change and interaction. In the words of Hindess and Hirst (1975), a "theory of modes of production involves abstract and general concepts – the concept of modes of production and social formation, of ideology and of politics of the state, and so on – but there is no general theory of modes of production" (p. 5). Even so, most theorists tend to see the power of global capitalism as so great as to dominate the process so that inequalities are created and tend to persist between capitalist powers and the pre-capitalist peripheries with which they articulate (Emmanuel 1972; Peet 1991).

For political ecologists, such a theory has many attractions. When Nietschmann's producers in Nicaragua, for example, encounter regional and global markets, we might be able to predict the inequalities in power and exchange that result from their "articulation" with a capitalist mode of production. Such imbalances can further be interpreted in terms of their changes in labor and exchange relations (reciprocity versus markets) and in the manner and mode of environmental extraction (craft fishing versus intensive harvesting of turtles), potentially leading to crisis. This approach expands common property theory appreciably, linking the process of accumulation with the encroachment and dismantling of traditional commons. Indeed, Marx's own argument on accumulation was predicated on his careful observation of the commons of the Scottish highlands appropriated by elites in the nineteenth century (Marx 1967a).

Even so, the role of the environment in human affairs remains somewhat vague in this formulation, since humans are portrayed as "promethean," capable of endless manipulation of natural systems as economies advance. It is unclear how the environment might influence history in such a general account. Other, more specific materialist efforts at analyzing ecological influences on the character and trajectory of society have been attempted, however.

The case of Oriental despotism

In perhaps the most prominent example, Karl Wittfogel argued that the roots of "Oriental despotism," epitomized by those Byzantine Asian bureaucratic states like historical China, lay in the establishment and maintenance of agriculture. Though a vociferous anti-communist theoretician, his thesis attracted the attention of Marxists and non-Marxists alike with its apparently far-reaching and explicitly materialist foundation. Writing in 1957, Wittfogel maintained, in classic materialist fashion, that the problems of production must be solved through political organization. Agricultural production systems in arid places, which depend heavily on big irrigation systems, must therefore require immense centralized bureaucracies. Thus, the political history of many of the world's great centralized states, particularly in Asia, can be explained as a simple result of the problems of water management.

This "hydraulic society" argument was attractive for a number of reasons. Self-perpetuating, totalitarian rule was a matter of political concern in the darkest days of the cold war and Wittfogel seemed to provide compelling scholarly explanation of global political structures. In this regard, the argument was especially attractive for Marxists, who held to materialist political philosophy but rejected the totalitarian communism of the Stalinist and Maoist states in the Soviet Union and China. How did utopian socialism turn into oppressive military bureaucracies? The apparent answer, Wittfogel suggests, is the deep historical roots of totalitarianism in "agromanagerialism." Materialism, in this way, provided an explanatory exculpation of Marxism; China is oppressive because of environmental history, not revolutionary precepts. The thesis was also of interest to materialist historians who had long puzzled over the different economic trajectories of the great empires of Asia and those of Europe. The West, with its rain-fed agriculture and smallholdings, by implication, developed feudalism and eventually private property, emerging as the triumphant capitalist center of the twentieth century.

Attractive as the thesis was, it was fundamentally flawed on both empirical and theoretical grounds. Empirically, detailed comparative and case-based research followed the publication of *Oriental Despotism* for the next several decades. The evidence, though sometimes contradictory, clearly demonstrated that large irrigation systems do not necessarily require central authority for management and can be managed by collective decentralized producer associations as effectively as centralized bureaucracies (Ostrom 1992). Furthermore, there is no evidence of an empirical association between large irrigation systems and centralized authority either in contemporary cases or ancient periods (Butzer 1976; Hunt 1988).

Theoretically, the hydraulic society thesis was also deeply flawed. *Oriental Despotism* follows a long tradition of *Orientalist* scholarship, which first assumes that "Eastern" civilizations are *essentially* different from those in the "West," and then proceeds to explain why that is so (Said 1978). In the process, the explanation serves to reinforce the artificial hierarchy of colonial thinking, where the traditionally colonized and dominated communities of the East (India, China, Arabia) are naturally inferior, bound in changeless tradition, and given to despotic rule. Such societies can only be changed and liberated by forces intruding from without – benevolently extending the enlightened advantages of the "West." Such colonial ways of thinking carry sweeping and problematic assumptions about the superiority of Euro-American society and culture. As Wittfogel insists: "nowhere, to our knowledge, did internal forces succeed in transforming any single-centered agromanagerial society into a multi-centered society of the Western type" (Wittfogel 1981, p. 227).

The hydraulic society thesis is, therefore, similar to many Eurocentric theories of history (Aston and Philpin 1985); it is both empirically untenable and politically problematic, since it ironically embraces and reinforces the very colonial hierarchies that critical scholarship claims to spurn (Blaut 2000).

Box 3.1 The intellectual politics of Wittfogel's *Oriental Despotism*

Karl Wittfogel's *Oriental Despotism* is an ambitious, frustrating, thoughtful, and vastly reductionist account of the roots and character of Asian civilization. In this book, subtitled "a comparative study of total power," he lays out his well-known carefully structured argument that despotic civilization in Asia, to which Soviet and Chinese communism are heir, is the result of traditions of bureaucratic control mandated by the management of large-scale irrigation systems. Despite wide-ranging empirical and theoretical flaws (Hunt 1988), the book is influential, especially amongst modern global economic historians.

Wittfogel was a scholar of tremendous intellectual and personal complexity. Associated with the Frankfurt school of social theorists, he drew upon the work of Marx (modes of production) and Max Weber (formation of bureaucracy) to analyze Asian society and history as an apolitical, scientific, and objective endeavor, as is evident in the tone and approach of *Oriental Despotism*. Wittfogel's career, however, bears testimony to the degree to which any scientific practice is rooted inevitably in a personal biography, enmeshed in broader political economy. As a refugee from Hitler's Germany, Wittfogel sailed to New York in 1934 to foster the Chinese History Project at Columbia University with the support of the Institute of Pacific Relations (IPR) and the Rockefeller Foundation. With the Nazi–Soviet pact of 1939, Wittfogel became disillusioned with Marxism, later moving to the University of Washington in 1949, where modernization studies centered on Asia were burgeoning during the cold war. As McCarthyist anti-communism heated up, such institutions became the target of government attention (Ulmen 1978).

When confronted with accusations that he was a communist sympathizer, Wittfogel was quick to denounce his colleagues and protest vociferously concerning his anti-communist credentials. Especially egregious was his denunciation of Owen Lattimore, a prominent Asianist, prolific and talented scholar, former friend, and fellow socialist. Under subpoena before the US Senate McCarran Committee, formed by Senator Joe McCarthy in 1950 and charged with investigating the IPR and other academic organizations that might be "subversive," Wittfogel repeatedly denounced Lattimore, insisting that he was not only subversive but, more importantly, naïve concerning the social and political character of Asian "feudalism." Lattimore, he charged, had been led by his subversive communist ideology to totally misanalyze the character of Chinese agrarian society and state power (Ulmen 1978, pp. 289–94). This was the beginning of an ongoing attack on Lattimore that ultimately ended in the destruction of his career (Newman 1992).

It was in the midst of this struggle that Wittfogel completed *Oriental Despotism*, his own treatise on Asian environment and power, which was finished in 1954 and published in 1957. Clearly *Oriental Despotism* and Wittfogel's other works are more than ambitious attempts at a comprehensive account of human–environment systems. They sit at the center of ideological and personal struggles that ripped at the core of academic freedom and political identity during the cold war. Wittfogel's journey, though marked by academic achievement, is also marred by personal and ethical tragedy, a testimony to the fact that apolitical ecology is ultimately impossible.

But the failure of Oriental despotism as a universal materialist thesis does not detract from the overall green materialist project. Even critics of the thesis admit to the importance of the linkage between production and the social systems of management they necessitate (Hunt 1988). Extensions of the broad argument to United States water management and the accumulation of bureaucratic agency power in the management of irrigation in the rural west, for example, has proved a productive line of thinking (Worster 1985).

The inadequacies of the hydraulic society thesis do provide, however, a cautionary tale for how political ecology must proceed as a way of explaining things. To avoid mistakes of reductionism, it needs to operate less from the universal and more from the particular, explore the context as well as the conditions of power, and eschew any simple narratives of social difference rooted in single-variable explanations. All the same, it must do so with a serious dedication to the material underpinnings of social life.

Dependency, accumulation, and degradation

Despite the shortcomings of some historical materialist research, materialist theory provided great explanatory purchase during the cold war period, and began to expose many of the more glaring sources of global inequality. Most compelling was the concept of *dependency*, first thrust onto the world stage by Latin American economists in the 1960s. For dependency theorists, the marginal conditions of the world's poorest nations were directly the result of the terms of trade established during the colonial period, when most colonized countries were forced to produce primary products, rather than more valuable industrial and craft goods. This was most notably the case in India, where a tradition of textile production was shunted aside by colonial authorities, who desired cheap cotton from Indian fields, but no competition in finished goods for textile mills in Manchester. These relationships hardened into a perpetual economic order of underdevelopment where, as Peet (1999) explains, "real power was exercised from external centers of command in dominant ('metropolitan') countries. Dependence continues into the present through international ownership of the region's most dynamic sectors, multinational corporate control over technology, and payments of royalties, interests, and profit" (p. 107). Even years after colonialism, and even where these poorer states are sovereign and control their own economies, their position in global trade remains disadvantaged as capital is accumulated elsewhere.

This holds implications for explaining ecological transformation in the contemporary world, and for exploring the relationship between economics and ecology. This linkage is built into materialism in a fundamental way since, as noted earlier, Marxist economics is based on the notion that capital accumulation requires the exploitation of both labor and nature.

For Marx, value comes from labor. Yet capitalists, he points out, make a handsome living without laboring in their own factories. The *surplus* – the difference between the value of the capital and labor put into a commodity (like a shoe, umbrella, or car) and the value accrued by the factory owner – must come from somewhere. The system of production under capitalism, Marx explains, is ordered so that workers, technicians, and engineers perform extra labor, the balance of which goes into the pocket of the owner, a non-worker. The same goes for nature; by expropriating nature's capital and underinvesting in restoration or repair of impacted ecological systems, capitalist firms squeeze surplus from the landscape, even and especially where commodity prices are falling and profit margins are tight. Moreover, the extraction of both labor and nature are simultaneous and interlinked. For crop production, for example: "all progress in capitalistic agriculture is the progress in the art, not only of robbing the laborer, but of robbing the soil; all progress in increasing the fertility of the soil for a given time, is a progress towards ruining the lasting sources of that fertility" (Marx 1967a, p. 506).

The same applies for forestry and other land uses. For environmental industries, the rate and intensity of extraction must always outpace that of restoration. "The development of culture and of industry in general has ever evinced itself in such energetic destruction of forests that everything done by it conversely for their preservation and restoration appears infinitesimal" (Marx 1967b, p. 248).

Tied to the concept of dependency, a pattern begins to emerge. Not only are the Miskito tied into a global economy where they are disempowered, they live within a Nicaraguan state system where they are most marginal, and where new demands for capital can only be met by exploiting their local natural resources, sea turtles. And though the Nicaraguan state at the time was ostensibly Marxist, it exists on the dependent periphery of a global exchange network, unable to establish favorable terms of trade. As accumulation continues, sea turtle overexploitation continues, social stratification increases, and the system becomes unstable.

Finally then, green materialism insists that such ongoing pillage of the environment must ultimately result in a political response. Just as the exploitation of labor leads to a labor movement, the exploitation of nature must result in an environmental movement (O'Connor 1996). In capitalism's excess, therefore, lie the seeds of more sustainable and equitable practices. The way these dynamics play themselves out in contemporary politics has, in recent years, gained a great deal of attention. Authors like Ted Benton (1996), John Bellamy Foster (2000), and James O'Conner (with the journal *Capitalism, Nature, Socialism*) all champion a materialist approach to contemporary environmental movements. This has arguably developed into its own distinct school of research into the *political economy of nature*, with work exploring the politics of water resources (Bakker 1999), of mining (Bridge 2000), and of training the environmental technocratic elite (Luke 1999).

Lessons from materialism: broadly defined political economy

The impact of this line of thinking on political ecology is somewhat more indirect. While not all of contemporary political ecology is explicit in its allegiance to materialism, much of the work at least tacitly assumes many materialist precepts. Among these, the most prominent assertions are that (1) social and cultural relationships are rooted in economic interactions amongst people and between people and non-human objects and systems, (2) exogenous imposition of unsustainable extractive regimes of accumulation result in environmental and social stress, and (3) production for the global market leads to contradictions and dependencies.

The degree to which these influences have either expanded or limited the scope of political ecology is a matter of debate. It cannot be disputed, however, that a form of materialism underpins much of political ecology and has motivated much of its research. Many and perhaps most practicing political ecologists would in no way identify themselves as Marxists or materialists. All the same, as the urgent questions of land degradation and human exploitation ignited the concerns of early research in the mid-1980s, materialist tools seemed to fit the bill quite well, and so continue to pervade the general consciousness of the field.

In this sense, almost all research in political ecology is theoretically engaged with what has often been described as a broadly defined political economy. The systems that govern use, overuse, degradation, and recovery of the environment are structured into a larger social engine, which revolves around the control of nature and labor (Althusser and Balibar 1970). No explanation of environmental change is complete, therefore, without serious attention to who profits from changes in control over resources, and without exploring who takes what from whom.

Even so, broad-scale materialist history and theory, attending to the articulation of modes of production and related concerns, did little to explain the specific and peculiar position of primary producers in the contemporary economy, especially local transformations of the environment that accompanied economic and political change. Nor did it provide much in the way of understanding rural primary producers – farmers, fishers, and herders – as important political actors, since most materialist theory focused solely on the potential power of urban working classes. Indeed, traditional Marxism was emphatic on the non-revolutionary irrelevance of the "vacillating and unstable" peasant classes and the disappearance of traditional peasantry into rural landlord entrepreneurs and wage workers (Lenin 1972a, p. 27; see also Lenin 1972b, especially chapter 2). Is the role of groups like the Miskito in history simply to become extinct or to wait on their liberation by urban masses? Materialism provided some tools for a nascent political ecology, but not the conceptual apparatus to understand the role of the daily life of producers in environmental history.

The Producer is the Agent of History: Peasant Studies

As it happened, the tools required to address the questions posed by the predicament of agrarian producer groups like the Miskito lay nearby. Investigation of these very dynamics – smallholder integration with broad markets, social unrest in rural areas, and political movements of agrarian communities in the face of coercive power – had become a locus of research activity in the social sciences in the 1960s and 1970s. The academic interest in these communities was not entirely innocent, since the cold war era saw a growing political urgency for understanding the world's poor agricultural communities. Fear of revolution made even very conservative thinkers interested in small, "backward," and agrarian places.

This is because revolutionary movements around the world were becoming increasingly rural in orientation. This turns traditional Marxian theory, where the urban working class was thought to be the vanguard of revolution and the peasants inert bourgeois conservatives, on its head. The Maoist revolution in China, leveraging the power of a massive peasantry, mystified both the capitalist West as well as

Box 3.2 Balancing the hatchet and the seed in Blaikie's *Political Economy of Soil Erosion*

As Piers Blaikie recently explained to me in reflecting on the intellectual context of the early 1980s, which informed his *Political Economy of Soil Erosion in Developing Countries* (Blaikie 1985), "I characterize neo-Marxist approaches of that time as a Soviet tractor – it produces loads of smoke, is sometimes heavy and cumbersome, but is still hugely powerful and does the work." The book expresses these very strengths and weaknesses. In a remarkably brief 150 pages, Blaikie provides a critique of neo-Malthusian explanations of soil erosion, unmasks the oversimplifications of technocratic solutions for complex ecological problems, and still has time to offer a sweeping theoretical account of what perpetuates rural soil erosion: capital accumulation by elite class interests. He further bluntly asserts that (1) soil erosion is only brought into check when it challenges systems of accumulation, and that (2) this doesn't happen very often.

With a background in geography with geomorphology and citing his influences as the radical development pragmatists Robert Chambers and Alain de Janvry, Blaikie was working in the 1980s, along with Nepali colleagues, to extend neo-Marxist thinking to serious environmental development problems. The initial monograph, "Centre and Periphery and Access in West Central Nepal: Approaches to Social and Spatial Inequality," was not widely published (maybe because of its title!) and so led to the later volume. The structuralist explanation that resulted is terrifically compelling and elegant. Filled with boxes, arrows, and flow charts that came to define explanation in political ecology, *Political Economy of Soil Erosion* graphically lays out causes and effects of erosion in Africa and Asia, showing that households make land-use decisions in broader economic contexts and that state policy in the postwar development era has made huge withdrawals from the soil bank of the rural poor to serve the interests of wealthier people in distant cities.

Twenty years later, Blaikie explains that *Political Economy of Soil Erosion* does not fit well with his current advisory and activist foci. Having moved to advocacy and paid policy work, radical critique of this kind has become more difficult, since "he who pays the piper calls the tune." *Political Economy of Soil Erosion* and an earlier book, *Nepal in Crisis* (Blaikie et al. 1980), both severely interrupted Blaikie's career as an international consultant, getting him temporarily banned from travel to Nepal, even while it did much to promote his career as an academic. Balancing criticism and effective policy intervention – weighing political ecology's hatchet against its seed – is demonstrably difficult.

Blaikie's relationship to Marxism is also marked by ambivalence. He explains:

> since PESE I remain a modernist, albeit a more modest one . . . a credible return to a structuralist approach now seems both implausible and undesirable . . . Anyhow, Marxism is thoroughly out of favor and is considered arcane and deeply flawed in most quarters. Also, PESE has since been criticized to have treated environmental politics rather cursorily, something I plead guilty to – it simply was not intended to be that kind of book. Still, my current interests in political ecology draw upon many Marxist ideas, although at a recent seminar I was roundly criticized for abandoning a classic Marxist approach and for letting the comrades down.

Soviet industrial Marxism (Wolf 1969). The Indonesian communist party (PKI), which would eventually lead a truncated revolution resulting in a bloody right-wing counter-coup in 1965 where a half-million people would be brutally massacred, was dominated by peasants (Cribb 1990). In the Americas, the prospect of revolution in the countryside was a vexing concern for both the authoritarian states of Latin America and for the United States, and a century of largely agrarian revolutions shook the *status quo* in Mexico (1912), El Salvador (1932), Bolivia (1952), Cuba (1962), and Nicaragua (1979).

Academically, however, interest in smallholding producers had been minimal and little was known about peasant behavior. The explosion of interest in the topic during the 1960s and 1970s, therefore, would pursue a number of previously unanswered questions. How do rural economies work? What levels of taxation and price control could they sustain? What leads to radical politics in rural areas? This broad field of investigation, made up of rural sociologists, anthropologists, political scientists, agricultural economists, and geographers, would come to be known for many years as peasant studies, spawning several academic journals with that title.

The term "peasant" has never been uncontentious. Often used as a pejorative term for smallholders – "rural cultivators practicing intensive, permanent, diversified agriculture on relatively small farms" (Netting 1993, p. 2) – "peasant" is a catch-all term. It stresses households that make their living from the land, partly integrated into broader-scale markets and partly rooted in subsistence production, with no wage workers, dependent on family and extended kin for farm labor. Peasants are autonomous in the sense that they are not solely dependent on markets for cash crops/products but they are also not isolated in the sense that they do interact with the broader market.

The knowledge and understanding of these kinds of communities that emerged during the early post-World War II development era was poorly informed and riddled with misconceptions. Peasants consistently frustrated the aggressive development efforts of proponents of green revolution technologies in the West and collectivism in the East. Why didn't the rural producer adopt modern methods? Why was it so difficult to organize them into large-scale collectives? For modernization theorists and Marxist planners alike, the peasant was a conundrum, and was considered to be (1) irrationally conservative, (2) inefficient, and (3) in a state of global decline, giving way to more modern farming arrangements. Peasant studies was to prove such assertions to be thoroughly unfounded and would insist that far from a minor or miscellaneous player in the global political economy, the peasant was an agent of history.

Chayanov and the rational producer

Findings on the peasant that contradict critical claims of development elites (in both the World Bank and the Kremlin) arguably begin with the work of A. V. Chayanov, an agricultural economist working in early-twentieth-century Russia during and after the revolution. As a key early player in Soviet planning, his views of the peasant put him at odds with Lenin and other leaders, who insisted that agriculture is more efficient when collectivized into large farms. The hesitation on the part of peasants to cooperate in these efforts, moreover, was seen to be a sign that peasants were inherently conservative petit bourgeois capitalists.

A populist pragmatist with a firm grasp on the logic of how farms work, Chayanov insisted that, on the contrary, peasants allocate labor not like a capitalist firm bent on capital profit, but instead with a thought towards meeting household subsistence needs while minimizing drudgery. They are risk averse but by no means disinterested in technological change, cautious but not irrational. Peasants are not inefficient at all, he argued, and they husband resources with great care, balancing household needs with access to markets for agricultural products and labor. In particular, the peasant household must be seen as one where the cost of labor is not calculated and where self-exploitation is central to survival and self-determination (Chayanov 1986).

While many of Chayanov's specific claims about farm behavior do not hold up in all places and times – his land-abundant context of Russia in 1920 is far different than the densely populated rice fields of Southeast Asia in 2003, for example – the basic approach and precepts do (Durrenberger and Tannenbaum 1979), driving a set of Chayanovian questions on into the present. Do peasants maximize profits or leisure? Do they minimize risk or drudgery? Do they more zealously defend integration or autonomy? And as export cropping replaces subsistence crops in some peasant households, should we predict accumulation and stability for the peasant household, or destitution? These questions would be the center of ongoing field research, debate, and modeling (de Janvry 1981; Ellis 1993).

Scott and the moral economy

While Chayanov provided an economic logic of smallholder production, he did not provide a political one (Brass 2000). Drawing on Chayanov, therefore, it is possible to make rough predictions about what a farm family with given resources might do with its labor during periods of scarcity or market fluctuations, for example, but it is not possible to predict when they might undertake political revolt, as the Miskito did. The intriguing possibility of an ecological theory of political resistance became the work of other thinkers, therefore.

Peasant studies provided two concepts and debates that would be crucial to political ecology in later years: *moral economy* and *everyday resistance*. These two ideas, the product of work by researchers E. P. Thompson and James Scott, help to form an image of the small producers who commonly play the role of protagonists in political ecological stories about land, forest, and fisheries.

The first concept – the moral economy of the peasant – holds that small producers are faced with subsistence risks that help to create social systems of mutual assistance and tolerable exploitation. Since, as noted above, peasants tend to be risk averse, they develop social arrangements and relationships that help them to redistribute surpluses and protect themselves in bad years. In a bad year, for example, a smallholder in Southeast Asia has traditionally been able to call upon reciprocal support from their neighbors or extended kin network, while in good years they can expect someone to make demands on their surplus grain. These systems are extended to shared land and labor as well as to the overall tolerance of smallholders for certain forms of redistributive rent, where a proportion of their harvest is lost in sharecropping and taxation.

The fundamental conclusion of work in this area is that some forms of extraction from peasants are acceptable to them (indeed they are moral subsistence *obligations*), while other forms are not. Peasants withstand exploitation, but not all kinds of exploitation, especially forms that put them consistently below a minimum line of subsistence, exposing them to undue risk. It is under the latter case that peasant movements arise to challenge political and economic authority. So too, as colonialism and market controls over household transactions increase, the moral economy becomes less stable, fomenting possible social upheaval (Scott 1976). Some have independently followed this line of thinking to suggest that the ferment of most contemporary revolution emanates from the countryside, where historically little sustained revolutionary momentum was visible (Wolf 1969).

Gramsci and peasant power

The second concept follows from the first. For rural producers, faced with increasingly exploitative relationships with local elites, outright armed resistance is often unfeasible or impossible, indeed it is unacceptably risky for most (Bowen 1986). On the other hand, ongoing *everyday resistance*, ranging from slander and back talk to work slowdowns and pilfering, can be used to oppose the limited social and ideological control of landlords and officials (Scott 1985b).

This approach to peasant resistance is rooted in the notion of *hegemony*, put forward by the Italian thinker Antonio Gramsci (1891–1937), a global traveler, critic, socialist, and activist. Sentenced to prison in 1926 and placed for the next decade in brutal solitary confinement, Gramsci wrote prolifically in a series of notebooks and letters before dying from a cerebral hemorrhage. Central to his concerns were both the coercive power of the state in its service to economic elites and the ability of the elite to achieve the spontaneous consent of the non-elite populous through the control of culture, opinion, and ideology. The limits of this coercive and cultural control, on the other hand, are the openings within which cultural and political resistance can occur, especially on a small scale, and especially amongst marginal populations, like peasants, without access to arms or more formal instruments of struggle.

Political ecologists would repeatedly acknowledge and use these ideas in research. As economic liberalization and market change occurs in, for example, rural grasslands of India, cooperative relationships begin to break down as the *moral economy* dissolves, potentially leading to overgrazing of the range (Jodha 1985, 1987). When the authorities in Madagascar restrict the setting of fires, an important traditional tool of local subsistence production, local smallholders respond in acts of *everyday resistance* by lighting more fires (Kull 2002). This cooperative, rational, risk-averse, authority-resisting, peasant is deeply embedded in the political ecological approach to explanation, which is often directed towards exculpating peasants from blame for land degradation and showing the adaptive social logics of cooperative, small-scale, smallholding producers and their daily acts of resistance.

Box 3.3 "Secret history" in Scott's *Weapons of the Weak*

Early in *Weapons of the Weak*, James Scott notes that intellectuals have a tendency to romanticize large-scale peasant uprisings and national wars of liberation (like Vietnam). The result of this fascination and concentration on "big" revolution is that scholars tend to miss the place where real resistance occurs: on the ground, every day, quietly, and with dogged persistence.

Drawing on his long research experience in Malaysia, Scott details the changes in the village of Sedaka through the turbulent green revolution period of the 1960s and 1970s, which brought large-scale irrigation schemes and mechanized double cropping to the rice-producing region. Like other political ecological accounts, there is a lot of discussion of emerging disparities between landlords and tenants and between farmers and farm laborers and of the troubling and oppressive conditions of a society in transition – paddy rice land is becoming scarce for the poorest farmers, wage losses have occurred in the wake of mechanization, and land rents are rising. Hard times for most, good times for a few.

But wait; all is not well in the orderly march towards class division and labor exploitation in Sedaka. As Scott expertly records, the poor refuse to be complicit in their oppression and go about resisting it at every turn. Workers fail to show up. Threshers carefully leave farmer's grain behind to glean later. Small livestock and fruits are stolen.

Wealthy landholders accuse their poorer neighbors of laziness and irresponsibility. Farmers, however, are not so easily fooled. Scott documents not only the methods by which poor farmers physically resist, but also the way in which they talk about and think about what is going on around them. This talking and thinking is an essential part of small-scale resistance, because it suggests solidarity and implicit organization, and shows that the dominant stories that empower the rich are not swallowed whole by those who work in their fields. Moreover, it demonstrates the way that political ecology, when executed well, can reveal the "secret history" (in Scott's terms) of political and environmental change by listening carefully to people on the ground. Documenting and making better-heard the sounds of such struggles might cause people in other places (workshops in India and grocery stores in Cleveland) to think about their own labor and their own relationship to power.

None of the small clashes in Sedaka ever boils over into full-scale uprising, however, since there is too much at stake on both sides to let it come to that. Instead, the rich and poor continue to snipe, pilfer, complain, and badger one another in an ongoing struggle. And this, it seems to me, is how class war mostly works, in barrooms, on shop floors, and in kitchens. As structures of inequality create intolerable conditions for working people, they consistently resist, back talk, and take what they can, fomenting class awareness in bits and pieces. Can such brush fires turn into sweeping change? Perhaps more quickly in Sedaka than in Wal-Mart, but the jury is still out.

All of these concepts are open to argument, however. Opponents have pointed to weaknesses in Scott's notion of hegemony, suggesting that his picture of top-down ideological control by elites (challenged in everyday acts of resistance) is an over-simple view of ideological control (Akram-Lodhi 1992). Others have argued that the peasant household struggle against risk is matched by an effort towards security from expropriation, and that cooperation comes from a coercive fear of loss rather than a shared moral commitment (Roeder 1984). The overall romanticization of local producers can also lead towards an under-appreciation of their stratification. Even so, after decades of scorn had been heaped on small producers, blaming them for everything from soil erosion and famine to deforestation and overfishing, political ecology would exhibit a refreshing enthusiasm for the peasant, echoing James Scott's sentiments when he urged "two cheers for the petty bourgeoisie!" (Scott 1985a).

More trenchantly, some have pointed out that by ignoring gender and the extraction of female labor value in the household and the difference between women and men in resisting authority, peasant studies approaches (especially those of Scott) provide skewed and unrealistic diagnoses of the village economic and social structure (Hart 1991). This latter criticism is part of a much larger and far ranging set of critical investigations that would also become fundamental to the formation of political ecology: feminist development studies.

Breaking Open the Household: Feminist Development Studies

In the post-World War II era, development assistance and investment swept the globe, led by large multilateral lending agencies like The World Bank Group – which includes the International Bank for Reconstruction and Development (IBRD) and the International Development Association (IDA) – and global superpowers like the US. Money poured into projects ranging from agricultural intensification to dam building and industrial development. By as early as the 1970s, however, it was clear not only that real economic growth throughout the "underdeveloped" world had not occurred but that, moreover, the position of many less powerful groups, especially women, had actually become worse in the process.

Agrarian development was especially problematic in this regard. Green revolution investments in farming had left women hungrier while advancements in rural processing and changes to cash cropping had made women poorer (Jain and Banerjee 1985; Soysa 1987). Critical feminist development theorists were quick to point out that these problems, though related to the historical marginality of women in rural areas, actually stemmed from the very development institutions that were designed to improve their lot. This included especially efforts in environmental development like forestry, farming, and water development (Shiva 1988).

As summarized by D. Rocheleau, B. Thomas-Slayter, and E. Wangari (Rocheleau et al. 1996), the central lesson from this apparently contradictory effect is that human–environmental interactions and processes are *gendered*, meaning that men and women experience the environment differently and often have different access to and control over ecological systems, as a result of their divergent social and cultural roles. Social conflict between men and women can be predicated on socio-ecological change, therefore, and changing gender roles or power can drive environmental transformation. This is true, they argue, in at least three ways.

First, women's survival skills around the world are based on a different knowledge of environmental processes and systems than that of men. Women in a rural African village may reject tree-planting schemes by well-meaning development officers, not because they are unaware of the value of forests, but rather because they know the selected species will fit poorly into local ecology. Moreover, women may depend on different species than men and utilize altogether different areas of the landscape (Fortmann 1996). State-imposed development schemes (crop improvement schemes, plantations, or breeding programs) or marketed products (seeds, trees, or hybrid cattle) for the improvement of production may take seriously men's environmental knowledge and priorities, while ignoring those of women altogether. This may lead not only to conflict, but to the collapse of environmental systems, tended to and managed by women (Cashman 1991).

Second, rights to access environmental resources are commonly differentiated by gender, as are responsibilities for the management of various ecological systems and functions. In many parts of Africa, land and tree rights are divided in a complex fashion, with gender figuring prominently in determining differential access. Gambian women, as a prominent example, traditionally control the products of all of their agricultural labor on private plots, but owe the harvest of communal village areas to the household. Development efforts that seek to alter local production systems with the goal of intensification may inadvertently reduce the resources that can be claimed by women, while increasing their labor burden (Carney and Watts 1991).

Finally, women are motivated to social and political action differently than men, *vis-à-vis* environmental problems and crises. Women under conditions of agrarian exploitation in Malaysia, for example, have taken action against their employers where men have not (Hart 1991). In a first-world urban example, community activism against the placement of hazardous processing in New York City's Harlem, like similar activism throughout the world, was led and championed by women of color (Miller et al. 1996). Efforts by men to expand collective agricultural production in Mexico, making heavy demands on female labor, have resulted in women's defiance (Mutersbaugh 1999).

It is important to note that the gendered differences in knowledge, access, and activism have little or nothing to do with physical/physiological differences between men and women – sex/biological differences. Rather, these divergent ecologies are the products of socially and culturally created structural positions relative to labor and nature. "Normal" women's work and "normal" men's work – what is expected of people based on their socially assigned gender – explains much about what women and men know about the environment, how much access they get to environmental systems, and their level of tolerance and resistance to environmental risks and burdens.

These three axes of potential gender difference hold implications for questions that concern political ecologists, explaining both environmental conflict and environmental change. The implementation of new high-intensity farming technologies like industrial fertilizers and pesticides in India, for example, may not lead to significant concern amongst male land managers in a rural setting. Indeed, the intensification of production may be welcome, since increased yields may put more money into the hands of producers. For women from the same households, however, the increase in cropping and extension of agricultural land actually results in an *increased* labor burden for women, who must travel farther and work harder to produce the same necessary goods that support the household, including fuel, fodder, and construction materials (Robbins 2002). The differential power of men and women and their uneven access to investment dollars from development may also lead to an expansion of cropping at the expense of lands for fuelwood, fodder, and medicinal plants (Robbins 2001). These conflicts take place within rural households rather than between them, and resistance to changes in ecological process might therefore be expected specifically from women, rather than from an undifferentiated peasant class.

Ironically, increased awareness of women's work resulted in an explosion of development programs directed specifically at women: the "Women in Development" (WID) approach of international aid. Busily trying to recruit women into development programs, either as recipients of aid or as "community participants" in a revolution of self-help, development activity unfortunately exacerbated many existing household tensions between men and women and, on some occasions, inadvertently reduced access to key resources upon which women depended. The resulting changes in human–environment interactions at household, community, and regional levels became a central focus for political ecology research, following its interest in understanding the unintended consequences of modern development (Jewitt and Kumar 2000).

But the underlying assumptions of some of this work also set traps for political ecologists trying to understand political and ecological change. In exploring the relationships between producers and environmental systems, there is sometimes a tendency to imagine that women are closer to nature and that their knowledge of the environment is not only different from that of men, but uniformly more accurate. These assumptions are problematic, especially when variations in women's experience are seriously considered: at different moments of their lives, under different socioeconomic constraints, and in varying cultural contexts around the world. Priorities and environmental knowledges of women vary tremendously, whether between wealthy and poor women in a Bulgarian village, between farming and herding women in Morocco, between white and black women near a hazardous New York sewage plant, or between women producers in an African peanut field and women consumers of peanut butter in a Canadian supermarket (Bonnard and Scherr 1994).

Even so, as with the other tools of critical theory that emerged in this period, the power of the approach is immediately evident. The hidden costs of agricultural change, for example, might now be seen simultaneously as costs in women's time, health, autonomy, and drudgery. Likewise, the decline of women's power and prominence in environmental management might be understood in terms of a concomitant decline in gender-specific expertise, critical for community survival. In the wake of feminist critique, labor, nature, and social power could now be linked through novel and powerful hypotheses and investigations.

Critical Environmental History

Yet like other critical approaches, and despite claims to an investigation of historical processes, the temporal depth of much of this work is remarkably thin. One might claim, for example, that current changes in planting and harvesting ignore women's *historically* evolving ecological knowledge or that changes in agricultural contracts lead to alterations in the *historically* flexible adaptive decision making of farmers, but these claims are commonly evaluated in only in *contemporary* development settings. History is theoretically important to political ecologists, but empirical research in these fields was initially not historical.

This had not been the case in cultural ecology, a field with direct methodological links to archaeology, which estimated pre-modern populations and measured ancient agricultural fields. The variables of political ecology, however, including social power, gender relations, division of labor, and economic structure, are somewhat less visible in the landscape and so harder to evaluate over the expanse of archaeological time.

In the humanities, however, concern about the environment has blossomed since the late 1960s, and the field of environmental history offers a powerful model for political ecologists interested in change over time. Pouring over the accounts of explorers, settlers, missionaries, business people, and administrators, environmental historians provide clues to long-term political ecological change. In France the tradition of weaving complex interlinkages between geography and history has long been practiced. The "Annales" school of history, established and championed by Fernand Braudel, documented the emergence of global economies in *longue durée* – history over the long haul – as focused around and through the environmental possibilities and limits of landscapes (see, for example, Braudel 1982).

The practice has since become established in Anglophone history. Donald Worster, in a prominent example, turned his attention to the American Dust Bowl, where drought and intensive farming methods together contributed to soil loss, blowing dust clouds, and the disruption of millions of rural lives, including bankruptcy and starvation during the 1930s. His work concludes that the ravages of the landscape were a nearly inescapable result of increased risk-taking farming behavior growing from capitalist agricultural economies established in the previous decades (Worster 1979). Caroline Merchant and William Cronon examined the histories of New England landscapes to trace the complex relationships between Indians and colonists, farmers and hunters, men and women, across the landscape of the colonial period. Their work points to the divergent ideologies as well as economies of settlers and Native Americans that influenced the peculiar ecological changes of the colonial period (Cronon 1983; Merchant 1989). Even more ambitiously, Alfred Crosby linked the historical waves of colonial and imperial migration and control to the expansion of "neo-European" ecologies, invasive plant, animal, and disease communities that followed and mutually supported European invasions of the Americas and Australia. Indeed, he argues, this "portmanteau" biota (so named for its coming along as part of conquest's baggage) was largely responsible for European colonial success in these areas (Crosby 1986).

This work holds several insights for critical environmental research. Firstly, historical research challenges the quick development "snapshots" of environmental research conducted in the present. Consider, for example, the degree to which Nietschmann holds integration of Nicaraguan Miskito ecology with larger markets to be the roots of contemporary crisis. What if that economic integration is really not so new, but instead dates to well before European contact? (McSweeney 2004). If regional and trans-regional exchange in native Central America is significant, can integration account for contemporary declines in turtle populations? Only multiple temporal scales of analysis can answer these kinds of questions.

Secondly, environmental history calls into question many of the discrete categories of environmental research. The environment, as traditionally considered by many researchers, is associated with rural areas away from the social and economic histories of cities. The bias of cultural ecology towards primary production systems (farming, ranching, and herding) further drives critical research into the countryside. Historical analysis has demonstrated the simultaneous development of cities and countryside, however, and explored their interlinked emergence over time (Cronon 1992). For a nascent political ecology, such an integrative view of the world is essential, if underdeveloped.

Thirdly, environmental history provides a powerful reminder that ecological change is not unidirectional. Despite the propensity of some environmental historians to tend to see degradation as a one-way path, landscapes over time usually show ongoing change, fluctuations, declines, and recoveries. Some of the seemingly wildest landscapes are the product of deliberate planning in the remote past (Spirn 1996), while many of the most apparently stable ecosystems have been subject to fluctuation in the past. Nietschmann's Nicaraguan turtle populations might have experienced wide-ranging growths and declines under varying climatic and oceanographic pressures over time. Historical analysis improves our understanding of just how precipitous the contemporary decline is, in a long-term context.

Despite these insights, environmental history has its limits. It is extremely difficult to compile thorough and comprehensive accounts of environmental conditions – comparable to those of contemporary ecological survey – from historical documents. Measures such as the diversity and ecological structure of a forest, or even the quantity of its canopy cover, are extremely hard to derive from narratives and, where they are available, are difficult to compare with contemporary data sources. This may lead to some unfounded or uncertain claims about important relationships in social and environmental change. Worster's *Dust Bowl* makes sweeping comments on the prehistoric state of grasslands ecologies, both in the United States and Africa, positing a long history of ecological balance against a contemporary situation of ruin. But what do we really know about grasslands ecosystems in Kansas before the arrival of capitalist farmers, let alone across the Sahel? Contemporary rangeland ecology suggests high levels of "natural" as well as anthropogenic variability in semi-arid regions, drawing such static and unidirectional models of change into question. Only through more exacting, small-scale, long-term, and detailed study of local ecologies, using the more exacting methods of historical ecology, can environmental history defend its often grand hypotheses (Russell 1997; Wynn 1997).

More significantly, the historical narratives are prefigured and limited by the larger contexts in which they are produced. Indeed, the writing of history is a political and social act, linked to, and embedded in, larger events and movements, including colonialism, imperialism, the cold war, and the contemporary struggles for global economic expansion and control. A re-evaluation of the role of history and science in politics has become the agenda of many theorists, whose influence on contemporary political ecology is equally profound.

Whose History and Science? Postcolonial Studies and Power/Knowledge

In his now classic volume *Orientalism*, Edward Said explained that specific forms of academic writing and analysis, in history, geography, religious studies, ethnography, and even economics, reflect scholarship that is enmeshed in the colonial and imperial contexts within which it is produced. Orientalism was that specific form of knowledge created to establish and describe the fundamental differences between something broadly defined as the East, characterized by mysterious backwardness and spirituality, and something called the West, characterized by forward-looking rationalism. Invariably, Orientalist narratives sought to explain why the East was one way and the West another. Said demonstrates that such writing did the colonial work of justifying the domination of one part of the earth by the other. In this way, Orientalist scholarship, by assuming difference, produced it in the process.

Orientalism, Said insists, has little or nothing to do with the Orient, *per se*. Neither does it reflect some simple, conscious, and cyclical nefarious "Western" imperialist plot to hold down the "Oriental" world. Rather, he suggests, Orientalist writing is a window into the minds, politics, and societies of its authors, linked closely to systems of political, social, and moral power that propel certain kinds of questions, descriptions, and answers, specifically in the context of American, British, and French scholarship in international contexts.

By way of example, in his historical and sociological analysis of India, Louis Dumont wrote that the Indian worldview was contained in a hierarchic ideology that "is directly contradicted by the egalitarian theory which *we* hold" (Dumont 1966, p. 4, my emphasis). The audience – "we" – is directly recognized in this passage, as throughout the rest of the volume, as civilized, egalitarian, democratic, and *Western*, living in freedom and thoroughly mystified by the hierarchic and structural world of "they." The balance of the text is a careful demonstration of the structural and hierarchic alien world of India. Backwardness, economic weakness, and dependency, by implication, are not causes of inequity throughout the region, they are its products. Leaving aside the ethnographic, empirical, and historical data that contradicts his thesis, the question that drives Dumont's narrative is in and of itself problematic. Following countless observers (from Weber to Marx), the thing to be explained is the Oriental "other," an object of mystery and fascination, whose shortcomings highlight the power and advancement of the author's own culture (Inden 1990).

Historically, colonial knowledges of the environment were constructed in a similar fashion. In the colonial administrative model, involving large-scale plantation, centralized irrigation authorities, and other modernization efforts, theories of environment were linked to theories of political domination. As environmental historian David Gilmartin puts it in the case of British colonial science: "the definition of the environment as a natural field to be dominated for productive use, and the definition of the British as a distinctive colonial ruling class over alien peoples, went hand in hand" (Gilmartin 1995, p. 211).

Contemporary apolitical ecology follows from much the same logic. For example, in his recent search for an environmental explanation of human civilization, *Guns, Germs, and Steel*, theorist Jared Diamond asks why *they* in the global south aren't as developed as *we* in the north are? Indeed, he more pointedly suggests that an "indigenous" New Guinean companion, Yali, posed the question, washing his hands of the clearly colonial implications of such a query and placing it in the mouth of the underdeveloped subject. His answer – the shape and latitude of northern continents determined the disparity – becomes more than a poor assemblage of empirical data, it is a Eurocentric history that can itself be seen as part of a long tradition.

More disturbingly, the neocolonialism of demographic apolitical ecology is evident in Malthusian views of development and global equity; they, the masses outside the lifeboat, are of a different order than we inside, making possible some remarkable pronouncements. "How can we help a foreign country to escape overpopulation?" Garret Hardin asks. "Clearly, the worst thing we can do is send food . . . Atomic bombs would be kinder" (quoted in Commoner 1988, p. 156). A human–environment science that begins with queries about *our* difference from *them* is ultimately a hand-me-down of not yet forgotten colonialism.

Indeed, the foundations of the most modern development projects arguably remain rooted in these same binary colonial logics. In the Orissan development case described previously, the introduction of optimal modern breeding stock was seen to require the slaughter of indigenous stock; the implementation of plantations meant the destruction of locally devised mixed land uses. These distinctions and conflations, dividing us/modern from them/primitive and local/backwards from foreign/progressive, stem from logics of domination, suggesting ideological controls of both environmental systems and local people.

The extension of this line of criticism against the binary logics of imperial science, along with a range of other critical examinations of "first world" science and humanities writing, has come to be known as postcolonial theory. The term is somewhat contentious. In one sense, it refers to a historical period – that of the contemporary post-colonized world, where unequal power relations prevail not only between colonial nations and their former colonies, but between northern academic writers, economists, and scientists and interpreters of the world in the global south. In another sense, it describes a methodology for approaching and investigating how European and American science is performed with specific attention to the context that influences its questions and answers (Mongia 1997). The challenge, postcolonial thinkers insist, is not only to explore and explain the dominant writings and theories about historically colonized peoples in terms of their contribution to global inequity and oppression, but also to rewrite history from the point of view of the colonized, rather than that of the colonizer.

This latter effort was codified by a group of historians (led by Ranajit Guha) who worked together to assemble a challenging and inverted picture of history, one where colonized subjects, often farmers, street dwellers, and workers in places like Kenya, India, and Indonesia, are active in global changes. Beginning originally from a South Asian perspective, this new view of the colonized subject, now referred to as the subaltern,* first sought to show that Indian independence was not achieved through the workings of educated middle-class nationalists, as it was commonly

* According to postcolonial theorist Gayatri Spivak, the term "subaltern" – originally meaning a lower-level military functionary – was introduced by Antonio Gramsci, who used the word to stand in for "proletarian" to escape the censorship of his writing during his long imprisonment (Spivak 1990).

characterized, but instead through the struggle and resistance of peasants and the urban poor (Chaturvedi 2000). This approach champions traditionally marginalized communities, therefore, by further implicating local elites in colonial efforts, showing how indigenous rulers and privileged classes were complicit in the domination of the truly poor. This approach overlaps heavily with peasant studies; drawing again on Gramsci, subaltern studies focuses on reclaiming marginalized communities to their place in history.

This postcolonial turn in the humanities and social sciences was predicated on several factors. First, since the 1960s there has been a great increase in the number of postcolonial "subjects" in the political and social realities of previously colonial states. Immigration has reversed the cultural tide of colonial influence. Even in Anglo-American academic departments around the world, including history, English, geography, and sociology, a broad range of active global voices are increasingly heard. Indeed, the unintended consequence of American and British historical sovereignty over Latin America, the Caribbean, and South Asia is the ironic increase in the number of scholars from those very places in the US and the UK. The intellectual dialogue that has resulted has fostered tremendous vigor in postcolonial debates.

Secondly, increasing political activity – or perhaps recognition of political activity – in local communities around the world is placing the voices of marginal groups squarely in the current debates on globalization and post-cold-war global governance. Instead of allowing loggers to clear traditional lands, indigenous forest dwellers respond by lying in the path of bulldozers. Rather than be displaced by massive dams in India, activists organize global-scale protests. States and firms have less and less choice about responding to groups they have long ignored. Thus the growth of postcolonial thinking mirrors the emergence of an active and global polity of historically marginal groups. The claims of development officials and academics are being directly brought into question by those who would be targeted for "development."

Thirdly, this change in thinking about the complicity of academic narratives in the extension of colonial power and repression, even narratives that ostensibly represent emancipatory ideologies like Marxism and feminism, is predicated on recent thinking about the nature of knowledge itself. This perspective, typically identified as *poststructural*, is one in which forms of knowledge can be explained by virtue of their relationship to establishing or subverting systems of power.

Power/knowledge

The term "poststructural" is somewhat too multifaceted and unevenly used to easily define here, but the power of the approach and its effect on contemporary social theory is significant. Associated with the writings of Michel Foucault, a former psychiatric worker turned historian–philosopher in the turbulent France of the 1960s, poststructuralism takes as its concern the instability of many of the categories we usually take for granted, including self, truth, and knowledge. Of Foucault's many influential theses, one of the most central was that truth is an effect of power, one that is formed through language and enforces social order by seeming intuitive or taken for granted. The key to understanding the character of society is to explore how certain taken-for-granted notions of the world are formed through *discourse* (language, stories, images, terminology) and how certain social systems and practices (medicine, forestry, prisons, schools) make them "true." By doing what Foucault referred to as archaeology – an effort to excavate the hidden history of meanings of concepts and things, along with their social and political histories – the hidden history of "truths" is demonstrated, making them appear less inevitable and showing their place in maintaining the power of individuals or groups. Some of Foucault's writings have been criticized for being opaque, but he summarizes this point fairly clearly: "Truth is a thing of this world: it is produced only by virtue of multiple forms of constraint. And it induces regular effects of power. Each society has its regime of truth, its general politics of truth: that is, the types of discourse that it accepts and makes function as true" (Foucault 1980, p. 131).

The implications of taking this notion seriously in environmental research are large. If accounts about people like herders or farmers or things like cattle or trees are conditioned and stabilized by social structures of power, the problem is not only understanding how social and environmental conditions change over time, or how they become undesirable, or how they can be changed. The problem is also understanding how scientific accounts, government documents, and local stories about those same social and environmental conditions are formed and made powerful by state institutions, media companies, experts, and families. How do specific ideas about nature and society limit and direct what is taken to be true and possible? For poststructuralists and postcolonial theorists such an investigation is a form of *deconstruction*, a word that basically means taking apart, undercutting, and questioning dominant truth claims. In particular, such an approach seeks to dethrone "hegemonic" discourses – those stories that hold a lock on the imaginations of the public, decision makers, planners, and scientists – so that other possibilities and realities are made possible (see Chapter 6).

Consider the explosion of pesticide use across American farms between 1945 and 1970. By the early 1960s this steady increase (to a level exceeding a half a billion pounds of active ingredient of pesticide per annum) had gone from an invisible and necessary practice to a serious and popularly perceived risk. With the publication of Rachel Carson's *Silent Spring* in 1962 (Carson 1962), researchers began to counter the apolitical explanatory orthodoxy explaining the proliferation of dangerous chlorinated hydrocarbons and organophosphates with counter-narratives of their own. Rather than accept the use of pesticides as an "acceptable risk" or as the "inevitable" price of progress, as advocated by the American Chemical Society (Marco et al. 1987), political ecologists began to ask why and how these stories and concepts were proliferated in the first place. How are notions of "development," "modernization," and "improvement" defined such that there is no alternative to high chemical inputs in agriculture? Deconstructive research must explore the history of ideas, examining the political effects, linkages, and sources of ostensibly "objective" and "apolitical" concepts, like "modern" agrarian methods, "improved" breeds, and "efficient" production.

Critical science, deconstruction, and ethics

In summarizing and justifying the goals of Orientalism, Said outlines the explicitly political character of research:

> the general liberal consensus that "true" knowledge is fundamentally non-political (and conversely, that overtly political knowledge is not "true" knowledge) obscures the highly if obscurely organized political circumstances obtaining when knowledge is produced. No one is helped in understanding this today when the adjective political is used as a label to discredit any work for daring to violate the protocol of pretended suprapolitical objectivity. (Said 1978, p. 10)

Taken together then, the importance for critical environmental research of the revelations from postcolonial theory, poststructuralism, and deconstruction for producing and examining knowledge and "truth" are twofold. First, such theory provides researchers with a powerful tool, perhaps one of the most important in the set used by political ecologists, for understanding the historical role of environmental science in the control of local populations and resources. When researching forestry in West Africa, for example, explanations can be offered for why colonial and postcolonial forestry officials insist that tree cover is declining when measures suggest they are in fact increasing; the discourse of degradation is a lever in power struggles between officials and local people (Fairhead and Leach 1995). Even where there is agreement that environmental degradation is occurring, as where ongoing soil loss presents a serious problem in Bolivia, in another example, the specific discourses that control and dominate debate as to its causes are central in establishing control of agrarian resources and politics (Zimmerer 1993).

On the other hand, the sharp hatchet of deconstruction cuts both ways. If scientific accounts of environmental change, including that of the political ecologist, are forged in the political context of discourse/power/knowledge, to what degree can the claims of critical environmental researchers, especially those from American/Anglo-European training, be viewed as an instrument of postcolonial hegemony and control? What might an archaeology of the apparently emancipatory theories of the political ecologist look like? How do we hear local voices if they are only mouthed through the foreign researcher? Can a non-indigenous observer effectively participate in an effort to write ecology from the point of view of the colonized? Is it right, or even desirable, that researchers play such a role? Are they even able to?

Postcolonial and deconstructive theory mandates an ethical evaluation even of what critical environmental researchers say and do. As we shall later see, this dilemma is not an easy one to resolve. Even so, joined to the several bodies of social thought informing research, knowledge/power approaches like these help to form the critical toolbox of an emergent political ecology.

Political Ecology Emergent

From all of this, it is possible to argue that political ecology as a distinct mode of research emerged in the last 20 years as a result of three convergent factors. First, cultural ecology and other related positivist human–environment social sciences had reached the limits of explanatory power for addressing some important questions about environmental change. Second, insights were emerging from critical theory of many kinds, including green materialism, peasant studies, postcolonial theory, and feminism (Figure 3.1).

Box 3.4 Revolution's influence "On the Poverty of Theory"

Michael Watts's 1983 essay "On the Poverty of Theory: Natural Hazards Research in Context" departs from just about everything. By arguing against "naïve empiricism," Watts departs from traditional geographic science. By deconstructing the neo-Darwinism of "adaptation," Watts departs from cultural ecology. By rejecting the "rational agent," Watts departs from the dominant approach to hazards research (Watts 1983a).

All of this departing must have taken a great deal of energy, but as Watts has explained, there was a lot going on at the time to fuel so much schism. Watts had joined with a group of "Nigerian and expatriate Left intellectuals" (in his own words) who were:

> working on similar problems rethinking postcolonial Nigeria – working and living in northern Nigeria during the oil boom of the 1970s . . . It had an urgency because of the ruinous famines in the region and because of a certain political energy around Third World revolutionary politics (as it then seemed) which included of course the revolutionary movements and successes of the 1970s (Vietnam, Nicaragua, Zimbabwe and so on).

Specifically, the project was to "denaturalize" things that were inherently social but constantly attributed to climate: famine, destitution, and poverty. The article sought to establish a critique of what Watts viewed as "naturalizing" approaches – empiricism, hazards, and cultural ecology – and the article formed the theoretical foundation of Watts's larger (indeed sprawling) book *Silent Violence* (1983b). This foundation was built on Marxism, peasant studies, and the British new left historiography of thinkers like E. P. Thompson, whose essay on *The Poverty of Theory* (1978) loaned its title to Watts's piece. But more directly, Watts drew on the thinking of Althusser and his structural account of class society operating like a big machine.

And to be sure, the essay is written in the language of high Marx. Typically, for example, the text urges that "the forces of and social relations of production constitute the unique starting point for human adaptation which is the appropriation and transformation of nature into material means of social reproduction" (p. 242). The often arcane prose (for those less initiated) can be frustrating, though it is crucial to his point.

But digging through the thickness of the language is well worth the effort, since the essay delivers a convincing story about the role of colonialism in shattering traditional responses to drought. It also provides an account of the horrors of famine formed from the paralyzing economic structures of emergent postcolonial states. All this and a critique of positivism in about thirty pages; not bad.

Most important, such structural critiques of famine cannot be mistaken as a fad of the 1970s and 1980s. The essay continues to be extremely influential, reflected in recent work on historical as well as contemporary famines (Davis 2001). Watts noted in 2002 that "I am perhaps embarrassed to say that I have not changed my ideas about famine and climate."

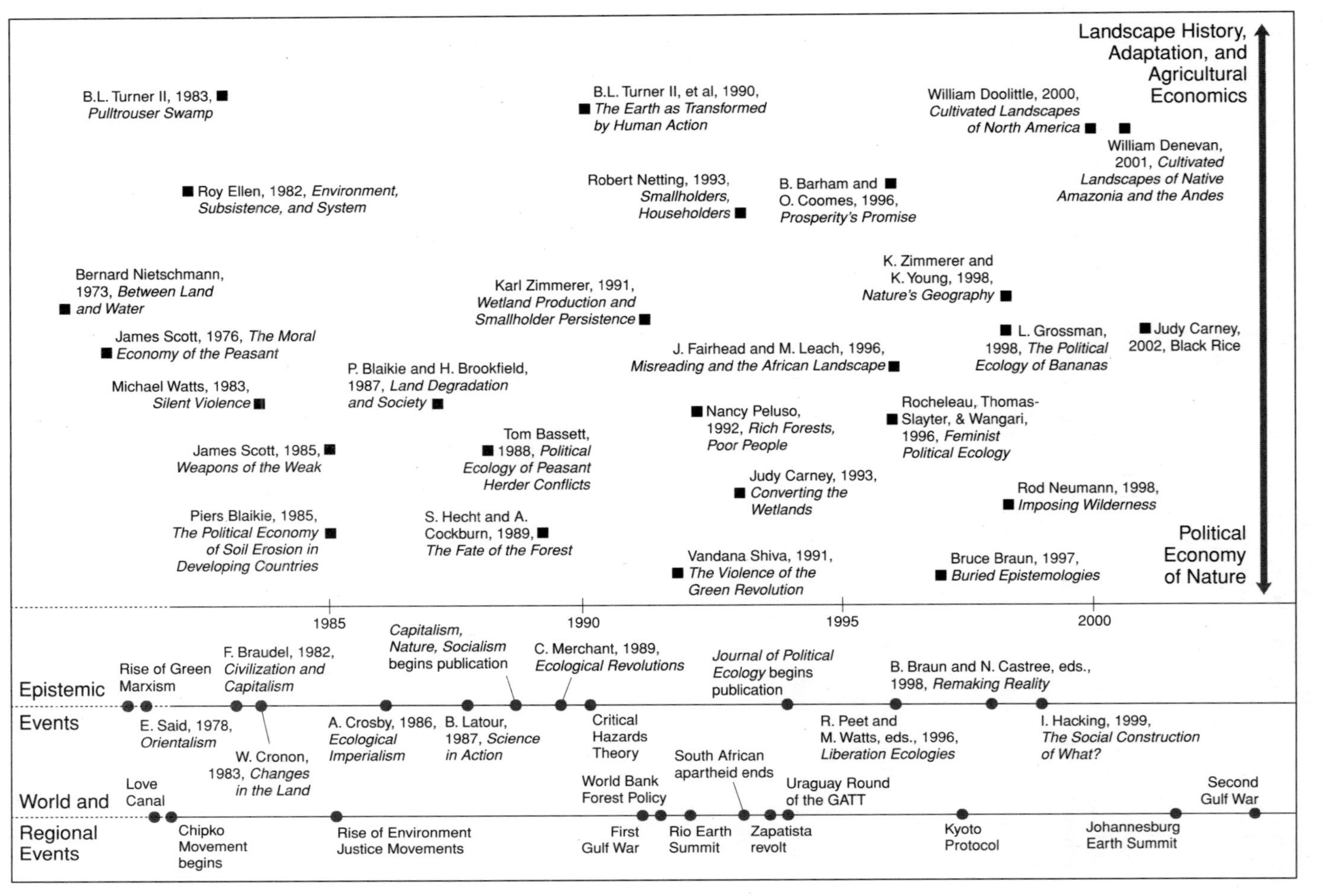

Figure 3.1 Integration and diversification

Third, apparent contradictions and feedbacks in global ecology were accelerating in the late twentieth century as a result of globalization. Images of Sahelian droughts were broadcast to households around the world. Global conservation organizations were beginning to vie for attention against multinational corporate machines. American consumers, while calling for preservation of tropical biodiversity, were eating bananas harvested from plantations that displaced rainforests. These trends were joined by peasant uprisings, disasters at Chernobyl and Bhopal, and popular movements in the forests on the periphery of the global economy. Thus while "apolitical" human–environment science continued to thrive as the twentieth century drew to a close, anthropology, geography, sociology, political science, and a gamut of other research fields were swept up in a great transformation, signaling a change in the methods of research and modes of explanation for social environmental science.

The excitement of that period for researchers in critical environmental studies should not be understated. At that moment, just as urgent problems appeared to be proliferating around the globe, a sophisticated mode of explanation was forming to explore the roots of such phenomena. Deforestation, soil erosion, and famines, long characterized as either "natural" and "inevitable" phenomena or the product of ignorant and overpopulated land managers, might now succumb to more systematic explanation. Political ecology had arrived, and at what must have seemed to be exactly the right moment.

4

A Field Crystallizes

In the wake of this theoretical activity, the scope of questions and processes to explore quickly expanded beyond the range of any single researcher, and the practice of political ecology began to vary as widely as its practitioners did. Some political ecologists continued to draw directly on the struggles and questions raised by thinking in peasant studies: How do subsistence producers respond to changing global economics? (Gupta 1998). Others focused on the imperatives of marginalization raised in postcolonial theory (Jewitt 1995). Some began to explore the character of landscape morphology under conditions of social and political struggle (Zimmerer 1991), while others investigated the social and political changes that grow from implementation of environmental control (Ribot 1996).

Concepts and processes in political ecology

Field/approach	*Concepts*	*Some processes to watch for*
Hazards	*Low and high risk behavior*	Traditional management systems, geared to minimize risk, are altered under political/economic pressure
Behavioral cultural ecology	*Rational land manager*	Production decision-making, geared to minimize drudgery, is altered under political/economic pressure
Common property	*Institutions as rules* *Collective action*	Dismantling environmental institutions in political economic change leads to system failure

Materialism/Marxism	*Surplus value* *Exploitation and hegemony*	Changing production systems increase exploitation and degradation of labor and environment
Peasant studies	*Moral economy* *Everyday resistance*	Reconfiguration of environmental management results in political and social crisis and resistance
Feminist development	*Division of labor and power*	Reconfiguration of environmental management leveraged on extraction of marginalized labor and resources
Environmental history	*Floating baselines* *Nature's agency*	New ecological systems emerge from competing and subsequent uses of the environment
Postcolonial/subaltern studies	*Political embeddedness of social science*	Accounts of social change used to extend and cement political control over marginal and colonized groups
Science and deconstruction	*Social embeddedness of physical science*	Accounts of environmental change used to obtain political control of people and resources

Despite this healthy diversity, the central concerns and questions of political ecology continue to revolve around several common conceptual tools and processes (see list above). These were codified in the first comprehensive text of political ecology, *Land Degradation and Society* (1987). Written by Harold Brookfield, a multidisciplinary cultural ecological pioneer, and Piers Blaikie, a critical development theorist, along with several other contributors, the book stakes out some wide terrain, defining many of the key concepts still used by political ecologists today. These include a cross-scale *chain of explanation*, a commitment to exploring *marginalized* communities, and the perspective of a *broadly defined political economy*.

Chains of Explanation

Placing local and regional environmental problems in a broader context has long been a core of environment–society research. When Humboldt describes the decline of the pearl beds of Cumana, pointing to the increasing rapacity of economics after the Colombian encounter (Chapter 2), he is contextualizing environmental change in political economy. He is, moreover, establishing a chain of explanation from pearl beds and their reproductive cycles, to producers and their fishing techniques, to markets and their demands, and finally to states and colonial powers, with their propensity for short-term benefits accrued in places distant from the site of extraction.

The cultural ecologist Andrew Vayda (1983) previously attempted to formalize this mode of explanation with the method of *progressive contextualization*. He suggested "we can start with the actions or interactions of individual living things and can proceed to put these into contexts that make actions or interactions intelligible by showing their place within complexes of causes and effects" (p. 270).

Box 4.1 The reluctant political ecologist: Harold Brookfield and *Land Degradation and Society*

Harold Brookfield does not lay claim to the title "political ecologist." But reflecting over several decades of human–environment work, he admits that he has actually done quite a lot of political ecology after all.

Brookfield began his career squarely in the "Pacific school" of cultural ecology, which was typified by long-term observation of the agrarian cultural practices of the people of Borneo and New Guinea (Brookfield 1962). Even as early as 1963, however, with publication of *Struggle for Land* (Brookfield and Brown 1963), a book exploring the problem of production and territory amongst the Chimbu of Papua, it is clear that Brookfield's concerns had inherent political dimensions. Later research projects on the effects of colonialism in Melanesia and the urbanization of villages in Malaysia were stamped with the effects of power and political economy in the lives of rural people.

It was not until the publication of *Land Degradation and Society* (Blaikie and Brookfield 1987), however, that these concerns came to be called "political ecology." Teamed in an unlikely way with Piers Blaikie who co-taught a joint workshop on land degradation with Brookfield in Canberra in 1984, Brookfield explicitly adopted some critical tools and *Land Degradation and Society* became a convenient rallying point for disparate research trajectories centered on the political economy of environmental change. The book had myriad contributors (often forgotten in the shorthand "Blaikie and Brookfield") including among many others the renowned common property theorist Narpat Jodha; it was in many ways a synthetic volume. Even so, Brookfield's voice (cautious, inductive, thorough) can be heard throughout.

Like most senior cultural ecologists, however, Brookfield remains wary of the Marxist and structuralist core concepts in political ecology, which were swept into the mainstream with the publication of *Land Degradation and Society*, although in a watered-down form. Like many other researchers, Brookfield fears that the introduction of a political economy approach lures thinking towards oversimple top-down explanation – or "structural determinism" in Brookfield's own terms.

Even so, the role of political economy is never far from his concerns. As recently as 2002, describing his experiences working in the United Nations University project on People, Land Management and Environmental Change (PLEC) in 12 countries from Peru to China, he concluded,

> not without some surprise, that there has been a good deal of political ecology in PLEC, and in the book on "agrodiversity" that I have written in the same period. But it has been a political ecology which focuses strongly on how individual farmers and their communities manage, adapt and innovate within the changing political and economic system, rather than on how the system determines what they do, and what happens to their land.

This suspicion is certainly understandable, since it reflects Brookfield's richly empirical and inductive approach to places, people, and systems. But reluctance to embrace the tools of critical theory was by no means passed along to future researchers. While Brookfield helped to build the doorway into a new mode of explanation, others have passed through it with far less trepidation.

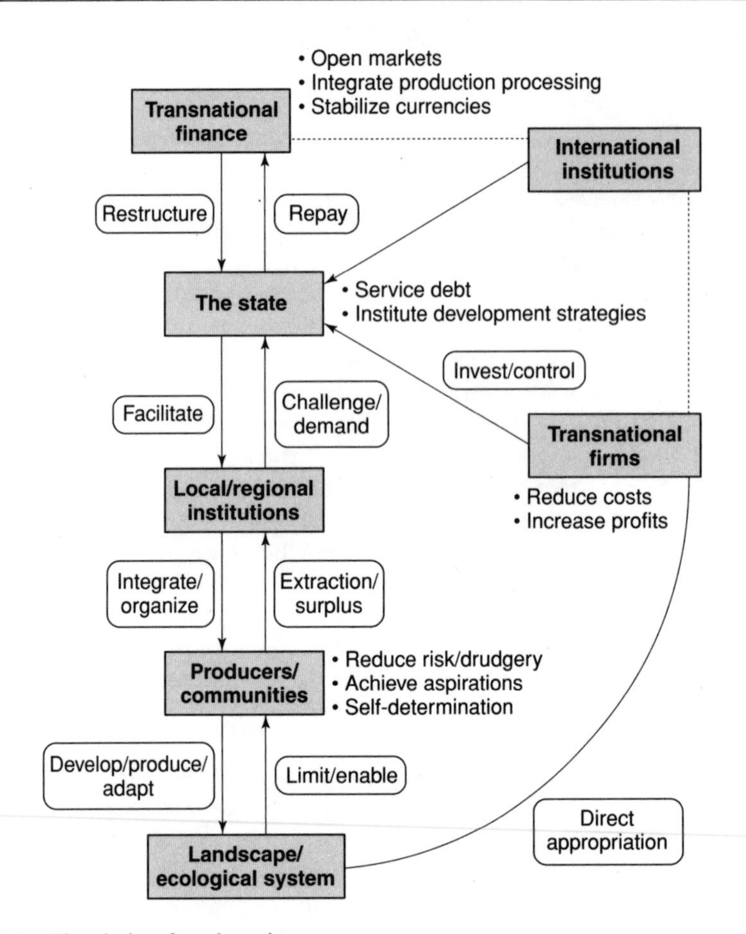

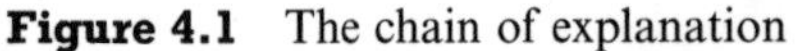

Figure 4.1 The chain of explanation

But Vayda's approach differs from that later offered by Blaikie and Brookfield, and that of political ecologists to follow, in at least one important respect. For Vayda, inductive observation can reveal all of the processes at work and determine all of the impinging factors on a "living thing," allowing an empirically accurate explanation of dynamics and change. Political ecologists, uneasily aware of the degree to which even simple and naïve induction tends to produce "apolitical" explanations that disguise the politics of the observer, are more explicit about the specific relationships for which they are hunting and that they expect to see. They are, as a result, far more candid about the role of theory in conditioning *what to look for* in explaining change: systems of extraction and exploitation, the role of power in determining conditions and context, the power of discourse to set the terms in which those contexts are produced. Blaikie and Brookfield explain that the chain of explanation:

> starts with the land managers and their direct relations with the land (crop rotations, fuelwood use, stocking densities, capital investment and so on). The next link concerns their relations with each other, other land users, and groups in the wider society who

> affect them in any way, which in turn determines land management. The state and the world economy constitute the last links in the chain. Clearly then, explanations will be highly conjectural, although relying on theoretical bases drawn from natural and social science. (Blaikie and Brookfield 1987, p. 27)

Of course, any given study will rarely include all of these elements, either because the regional context renders them irrelevant or because a single specific link is of greatest importance to the researcher. Research may, for example, focus solely on the way state authorities respond to elites and economic pressures to determine the regulation of specific forms of extraction (e.g., clear-cutting, strip-mining). Other research in the same context, on the other hand, might just as easily focus on the way producer communities respond differently to altered institutional arrangements, mobilizing different resources to adapt their use of the landscape with varying ecological outcomes.

The problem in assembling such explanations is that selecting the suite of variables and the appropriate scale is difficult and must be driven at least in part by theory. The chain of explanation is as much art as science: "The knack in explanation must lie in the ability to grasp a few strategic variables that both relate closely together in a causal manner, and which are relatively sensitive to change. In that way the most promising policy variables and paths to social change can be identified" (Blaikie and Brookfield 1987, p. 48).

Peanuts and poverty in Niger

Early political ecology on the causes and impacts of drought and famine in Niger in the 1970s is illustrative. This crisis occurred within the larger context of the Sahelian famine, which between 1968 and 1974 left thousands of people dead, half of the regional livestock herd depleted, most of the total crop ruined, herders and farmers in bloody conflict, and global decision-makers scratching their heads for answers. Sub-Saharan Africa, though having a predominantly agrarian population at the time, and having long been a food exporter, became a net food importer during this period. Production – the amount of food produced per unit of land – also fell dramatically over the period, despite efforts in technical development to raise yields. Soil loss on farm and pasturelands became an acute problem, with bare and exposed lands subject to wind erosion. The crisis persisted into the 1980s, leaving many regional economies and ecologies in disarray. The promulgation of advice and recovery plans perversely became something of an industry in itself, with bad data recycled to suit a range of theories, and little research on the ground (Watts 1983b).

As international analysis drifted between overpopulation and technological backwardness, pioneering political ecologists in the 1980s turned to a more robust empirical and theoretical exploration of the ultimate causes of crisis. In the process, they set the precedent for how political ecology is practiced.

What made Niger and other countries in the region so vulnerable to drought? Breaking the global silence, anthropologist Richard Franke and sociologist Barbara Chasin conducted a series of studies on the question. Their answer came in the form of peanuts (Franke and Chasin 1979, 1980).

Peanuts had been introduced to Niger in the last century and championed both by late-colonial agricultural officers and their post-independence counterparts as a food oil source that might compete in global markets, especially with the growing North American soybean crop. Developing Sahelian nations were to earn export income from burgeoning harvests while European processors would receive cheap inputs. Everyone would win (Franke and Chasin 1979, 1980).

Yet the local ecology of the Sahel was ill-suited for intensive peanut production, at least as a reliable income source. So too, the global oil markets were headed in the opposite direction of development plans for the region. Prices paid to farmers, originally inflated through subsidies, fell by 22 percent in the late 1960s, just prior to the drought. The results were devastating. The imbalance in the economy towards a single export crop – in 1970, 65 percent of the country's total exports came from peanuts – scaled up the ecological failure of the peanut to a national level. The decline in fallow time and intensity of cropping demanded by the crop resulted in dramatic soil nutrient declines and moisture loss. The drought and its accompanying winds carried away much of the topsoil around the failed crops. The expansion of cropland for the peanut, moreover, pushed nomadic (Fulani) herders out of the region, northward into increasingly fragile and variable arid biomes. The resulting herd losses, especially with the onset of the drought, were disastrous (Franke and Chasin 1980).

In apoltical ecology, famine, soil erosion, economic crisis, and starvation during the period might be viewed as a problem of overpopulation, or the random effects of the environment, or perhaps an irrational resistance on the part of local farmers to utilize modern technology. But by contextualizing farm, herd, and soil failures along a chain of explanation, a somewhat more deeply structured problem emerges.

Marginalization

Along with the chain of explanation, political ecology brought with it a theory of feedbacks in socio-environmental change, expressed in the concept of *marginalization*. Blending concepts of the margin from neo-classical economics, ecology, and political economy together, Blaikie and Brookfield (1987) offered marginalization as a process that led to simultaneous and increasing impoverishment and land degradation in and amongst the global poor.

In neo-classical economics, the concept of margin denotes the limits of production, where increasing effort, labor, cropping, and cutting in the landscape provides less output per unit of input. Consider, for example, an agricultural field where increased intensity, through the addition of more and more inputs, continues to increase the amount of crop returned, but does so at a falling rate per unit of pesticide, fertilizer, or labor.

From ecology, the concept of the margin suggests ecosystems that are either unstable or vulnerable to change or slow to recover from disturbance. Consider the difference between a pasture on deep rich soil in a tree-sheltered area of low slope versus one on a thinly covered stretch of steep open ground. The same rainfall event or windstorm will have a very different effect on each, with the former being less affected and recovering its productivity quickly while the latter, more marginal landscape, is more seriously affected and slow to recover.

From political economy, marginal communities are those at the fringes of social power, with little bargaining strength in the market and little force in political process. Marginal people, ranging from women in some regions and times to lower classes, castes, tribal groups, and minority ethnic communities, are often excluded from labor opportunities, important resources, or control over decision-making.

Marginalization, Blaikie and Brookfield argued, is a process whereby politically and socially marginal (disempowered) people are pushed into ecologically marginal (vulnerable and unstable) spaces and economically marginal (dependent and narrowly adaptable) social positions, resulting in their increasing demands on the marginal (increasingly limited) productivity of ecosystems. As a consequence, those individuals and groups will tend to increase their efforts on the landscape, increasingly pushing the limits of its capacity, and achieving lower and lower yields. The result is a degraded landscape that returns less and less to an increasingly impoverished and desperate community – a cycle of social and environmental degradation.

The Silent Violence *of famine in Nigeria*

For those trained to think about the adaptive capacity of people and the incredible resilience of communities that live in uncertain environments, the concept of marginalization proved useful. As the Sahelian crisis continued through the 1970s and 1980s, more questions were raised. Weren't the farmers and the herders that lived there already well adapted to the common problem of drought? And if so, why were those adaptations failing in the contemporary period?

An answer came in the exhaustive study of Nigerian famine, *Silent Violence*, by Michael Watts, a critical geographer trained in hazards and cultural ecology, who connected the crisis in the Sahel to a century of economic and social transition. The problem lay, Watts argued, in "the rupture of local systems as they become part of coherent and highly integrated global networks" (Watts 1983b, p. 14). Following traditional cultural ecology, Watts's work uses intensive field methods to show the strategies of traditional producer response to rainfall and market variations. Added to this, however, is a carefully constructed account of the flow of capital and debt in the colonial and postcolonial periods, culminating in subsistence crisis for otherwise traditionally well-adapted communities. The resulting 500 plus pages of dense text in *Silent Violence* reveal (in often somewhat difficult language) the ecological and economic feedbacks of marginalization, where decreasing social power and resource access leads to depletion of resources, reinforcing social and political subjugation. Moreover, they serve to remind us that while drought is a *climatic* event, famine and mass starvation are *social* ones.

While shortages occurred in pre-capitalist economies, access to food in the past was determined by systems of reciprocity and social exchange, as suggested by Scott and other "moral economists" (see Chapter 3). Following this logic, Watts argues that famine in the historic past, especially in the precolonial village/state economies of Hausaland and the Sokoto caliphate, was held at bay by social/environmental institutions. Historically, risk-averse peasant households in the region carefully mixed crops, keeping drought-tolerant, less productive, and less valuable crops mixed with higher output and value species. Stockpiling and redistribution at the community level through

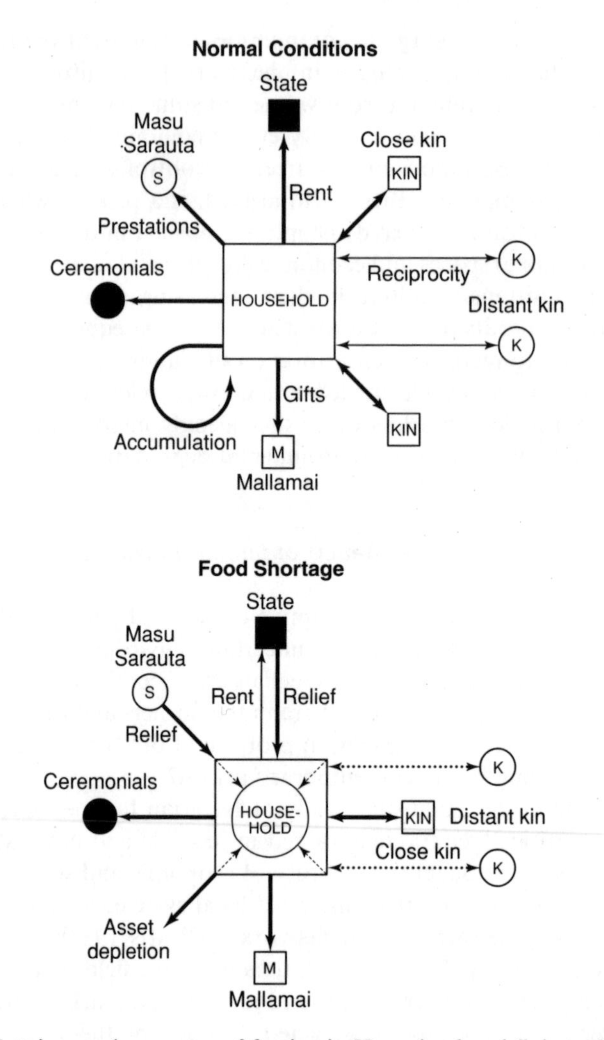

Figure 4.2 Watts's moral economy of famine in Hausaland and Sokoto (Reproduced from Watts 1983b, figure 3.2)

mandatory social gifts, ceremonial offerings, and other exchanges, further offset dry-year scarcities. Famine was deterred also through a state-structured system of rents, which allowed redistribution of surplus grain during famine periods. During normal (rainfall) conditions, households paid rents to the state, exchanged extra grain with kin, and accumulated and paid out small surpluses as a reserve against famine (Watts 1983b).

Far from a romantic or ideal cooperative community, the regional states of this period are shown to be slave-based, exploitative, and highly stratified. This was by no means a happy and cooperative world of selfless "mutual aid." Even so, rooted in an Islamic moral ideology and tied to a carefully articulated code of obligations and restrictions, this social system averted severe famine, and held in place a series of graded and institutionalized responses under conditions of worsening drought, which made crisis situations predictable and manageable.

Tracing the rise of economies from the colonial merchant period to the post-independence development era, *Silent Violence* demonstrates the way in which capitalization of exchange relations transformed famine ecology. Seeking petroleum receipts and cash-crop exports, the Nigerian state subsidized programs for cheap food imports and industrial development. As more household and farming inputs were purchased and more crops grown for distant markets, producers were driven into higher-risk decision situations. The concomitant increase in rural debt was coupled with the disappearance of redistributive systems of previous regimes. The results were cropping decisions that made increasing demands on soil and water and which left producers open to the vague and turbulent roller-coaster ride of commodity market price changes. Many landowners were at the same time forced into wage labor, allowing less time and labor for farm work and land improvement in the most marginal households, leading to soil degradation.

With the arrival of a near-record drought in the 1970s, this highly risky and turbulent ecological economy was primed for failure. As increasing income inequality drove more households into increasingly marginal conditions, the effects of the drought and its feedback onto the land became more intense. And even though increasing petroleum revenues were pouring into the country, the Nigerian state used the countryside as a cash crop earner of export income, even while investing little or nothing in its maintenance or improvement. Nigeria was a political ecological disaster (Watts 1983b).

Broadly Defined Political Economy

Land Degradation and Society also codified the notion that compelling and robust explanation of social/ecological process and change requires attention to political economy, broadly defined. Decisions by land managers are predicated on the economic pressures that determine management goals and the political contingencies that determine management opportunities. The same herder, for example, will make very different decisions when meat prices are low and access to land is high than when the reverse is true. Changing political and economic conditions therefore alter the context of decision-makers and set the terms for their use of the environment.

Narrowly defined political economy concerns itself with labor and ownership relationships. Formal materialist/Marxist analysis of political economy usually directs attention to a specific set of processes, especially the transition to capitalist systems of production where non-workers accumulate capital and land, while historically independent or self-sufficient producers are forced to sell their labor directly in the form of wage work. But even where control of environments and of decision-making is not a question of capitalization and markets, political economy is useful nonetheless. In pre-capitalist New Guinea, for example, the agronomic system of sweet potato production changed over time to a specialized system of gendered labor where men prepared land, accruing capital benefits, while women weeded and harvested. A political economic trend towards differential household power followed, and set the conditions under which each group might differentially respond to technological innovation and intensification (Allen and Crittenden 1987). This is a political economy, though not necessarily one related to a cash economy.

In this way, almost all political and environmental explanations center on who controls resources and how the rules and conditions of production and exchange are set in political struggle. But this political economy is defined very broadly to encompass a range of spheres in which power is exerted, whether it is control of labor, land, or ideas.

Struggles in Cote D'Ivoire

These concepts again proved useful for early political ecology. In the years following the crisis, conflicts began to break out between peasants and herders in the northern parts of the small West African nation of Cote D'Ivoire. During the early and mid-1980s increasingly hostile encounters began to occur between Senufo farmers and Fulani herders, which led to cattle theft, assault, arson, and murder in the region. Typically, such conflicts are treated in the global press apolitically, either as ancient ethnic hatreds or as problems of scarcity – too little land, too many animals. But Senufo–Fulani relationships are historically cooperative and as geographer Tom Bassett put it, "this dwindling land base model does not adequately explain why and how some groups gain access to land while others are losing their rights. Nor does this perspective explain why conflicts exist between peasants and herders in relatively land abundant areas like northern Ivory Coast and central Camaroon" (Bassett 1988, p. 453).

Rather than view the problem as a locally bounded pressure/response system, Bassett suggested that "ultimate causes" lay behind the conflicts, located in the development economics of regional and global systems of extraction and growth. Specifically, Bassett argued, following several years of fieldwork in the region, the conflict is a result of crop damage caused by herds that have moved into the region under a carefully supervised national and international development scheme.

The Ivorian state, like others in the region, found itself in a fiscal crisis precipitated by debt payments, suspensions, and austerity measures imposed by the International Monetary Fund. In order to earn enough capital to finance the massive development loans of earlier decades, Cote D'Ivoire established a range of "parastatal" joint ventures, in which foreign investors from the US, Japan, and Europe expand agro-industries in collaboration with the state. Rural producers, in particular, are taxed heavily, either through direct taxation or through price controls, in order to subsidize urban industrial development (Bassett 1988).

As a strategy to further such export income, the Ivorian state established a livestock parastatal, SODEPRA (Société pour le dévelopment de la production animale). The firm was designed to attract foreign herds into the country for meat processing by distributing grazing rights to Fulani migrant herders moving south out of the more arid Sahel. Though the Fulani had migrated into Cote D'Ivoire seasonally or during low-rainfall years in the past, the massive arrival of large herds from Mali and Burkina Faso in the SODEPRA plan was coupled with "pastoral zone," "micro-zone," and sedentization schemes that established Fulani grazing in densely populated agricultural areas. The resulting crop damage, where cattle entered and grazed in Senufo fields, occurred on a regional scale, decreasing local yields. Farmers responded directly by stealing cattle. At the same time, however, farmers seeking to capitalize on manure-

Box 4.2 Putting the geography back: Bassett's "Political Ecology of Peasant–Herder Conflicts"

Tom Bassett's "Political Ecology of Peasant–Herder Conflicts in the Northern Ivory Coast" (1988) is the first article to refer to itself as "political ecology" in the flagship journal of the discipline of American geography – *Annals of the Association of American Geographers*. In it, Bassett argues that peasant–herder conflicts come not from ancient hatreds or absolute resource scarcity, but instead from ecological contradictions created by the state's economic planning. As much as for its timeliness, the essay was important to political ecology because it sought not to explain ecology with politics, as laid out in Blaikie and Brookfield's almost simultaneous publication of *Land Degradation and Society*, but rather to explain politics with ecology.

The argument is straightforward but the evidence is multifarious, largely because of the complexity of local livelihood systems and their interrelationships. Bassett argues that the growing tension between Senufo farmers and Fulani herders developed from crop damage from herds in high densities entering agricultural zones. These encounters grew from state schemes to draw pastoral people and their herds into the country. These were, in turn, directed to help develop an export-earning meat economy, which in turn was designed to offset massive debt. Debt, state strategy, settlement, crop damage, violence: a classic chain of explanation.

The motivating force behind making such a profoundly structural argument was an impatience with the fuzzy atheoretical and descriptive human ecology that Bassett saw in the research around him. As he explained in 2002, "I felt dissatisfied with the descriptive 'check list' approach . . . geographers typically presented a non-theorized list of socio-economic and cultural factors to explain resource use patterns."

At the same time, however, the work represents an effort to put geography (seasonality, rangeland condition, cropping cycles, herd mobility, burning, and cattle disease) back into discussions in anthropology and peasant studies that had focused less on these kinds of material conditions. By uniting these issues with the concerns of political economy (the state, foreign capital, agricultural policies, surplus appropriation, poverty), Bassett says that he "hoped to both explain the nature of peasant–herder conflicts and point to political ecology as a viable geographical approach to development studies."

Curiously, Bassett first chose to use the term "political ecology" to describe this work without having seen the phrase before, and without having read Blaikie and Brookfield's key text. This suggests the degree to which the questions that make up "political ecology" were permeating lots of people's thinking at that time, and underlines how the field was a part of a larger shift in academic thinking about nature and culture.

rich areas around Fulani corrals often encroached on herder lands around Fulani camps. The resulting conflicts became more acute annually (Bassett 1988).

Ultimately then, farmer anger and frustration stems from the instability of their land tenure rights, essentially condoned by the state, coupled with the loss in crop productivity from herd activity. Fulani on the other hand, lured to Cote D'Ivoire with promise of land rights, find themselves with tenuous claims to land and at precarious risk from grazing limits and farmer retribution. The damage to Senufo crops and cattle thefts from Fulani herds, therefore, is an indirect form of uncompensated capital expropriation, upwards to parastatals, and outwards to the cycles of circulating global capital. The violence that results occurs only between marginal rural communities, even while the pressures that drive the process are structured to benefit an indigenous and foreign business elite, living in the cities, far from harm's way. In this sense, political economic dynamics are fundamental to conflict in the region, but only if defined broadly enough to incorporate inter-ethnic dispute, parastatal incentives, and the unremitted value of crops.

While the political ecology of sub-Saharan Africa would mature considerably in the next decade, unhinging some of these findings (Batterbury 2001; Batterbury and Warren 2001; Warren et al. 2001), these early efforts established many of the explanatory patterns and modes of analysis that would become typical of political ecology. More importantly, they confronted localized, racialized, and demographic apolitical ecology with a radical alternative. Regional environmental problems were the systematic result of marginalization, exploitation, and the struggle for control of productive resources.

Twenty-five Years Later

A quarter of a century later, these themes remain important and relevant, but new issues have emerged and the landscape of global ecological struggle has changed. Work in political ecology has been redirected as a result (Table 4.1).

In the last decade, for example, nationally established conservation areas have increased almost fourfold, and the land allotted to these areas, as well as to international biosphere reserves, has increased dramatically. Indigenous people occupy

Table 4.1 New ecologies, new politics

Environmental and economic indicators	*1990*	*2000*
Number of national protected areas[a]	6,931	28,442
National protected areas or biosphere reserves (millions of hectares)[a]	803	1,115
Total forest (millions of hectares)[a]	3,593	3,454 (1995)
Plantation forest (millions of hectares)[a]	28.83	68.35
Debt of low- and middle-income nations (billions of US dollars)[b]	1,460	2,573
Percentage of female workforce in agriculture in low-income nations[b]	—	74
Fertilizer use in low-income nations (100g/ha of arable land)[b]	290 (1979–81)	632 (1996–8)
Food as percentage of merchandise imports in low-income nations[b]	7	14

Source: [a] World Resources Institute 1992, 2001
[b] World Bank Group 2001, http://www.worldbank.org/data/databytopic/databytopic.html

many of these lands. This raises the question not only of what political economic forces create biodiversity decline, but more trenchantly, what are the politics and economics of enclosing lands belonging to the poor in the name of conservation. Who benefits and who loses in the preservation of biodiversity?

Total forest cover is decreasing globally, hinting at the political forces of deforestation described previously. At the same time, however, the amount of newly created forest has nearly doubled, suggesting that the new political ecology of forestry is not so much about decline and scarcity, but more about what kinds of ecologies are being produced, by whom, and to whose benefit. Who defines what a forest is and who controls its location, expansion, and growth?

While the debt of low-income nations continues to expand rapidly, the proportion of food imports to total imports in these nations is also on the rise. This raises questions not only about the possibility of starvation accompanying fiscal crisis, but also about the pressures on regional production for agricultural and pastoral producers competing on global markets. Can formerly self-sufficient producers earn enough from cash crops to feed themselves? At what cost to the soil?

Fertilizer use in low-income nations has more than doubled in the last 20 years. This suggests not only that producers are dependent on expensive inputs to sustain agrarian livelihoods, but also that the soils of the underdeveloped world are increasingly exhausted in the intensification that chases declining margins. How have traditional nitrogen cycles been broken? Can they be re-established?

Finally, in the low-income nations of the world, the proportion of the female workforce in agriculture is 74 percent, relative to 62 percent of the male workforce. As burdens of debt, state restructuring, and capitalization are played out as environmental transformations in the countryside, it is increasingly the knowledge, actions, and adaptations of women that will determine the political ecological conditions of the underdeveloped world. How does differential gendered access and control of resources affect the way women cope with political and ecological change? In what ways does this direct their transformation of the environment?

These snapshots of changing global conditions hint at a number of compelling and urgent questions. These questions, joined to more traditional concerns, have become the expanded purview of political ecology in the last 25 years. While research still asserts linkages between regional environmental systems (soil, grasslands, water, biodiversity) and political and economic process (debt, land tenure, and access), it is concerned increasingly with themes beyond the scope of that earlier work, including: producer livelihood adaptation and diversity; community and participatory development; state and international environmental conservation; intra-household divisions of labor and resource access; social versus official, state, or scientific knowledge; and social movements and group conflicts in resource access.

The pursuit of these themes has in turn given rise to broad common areas of research and argument, which encompass a diversity of approaches. All of these research foci, however, depend upon defining, identifying, and measuring ecologies and environments. Research, therefore, is commonly predicated upon determining either the material condition of the environment (e.g., soil conditions, land cover types, or groundwater levels) or its imaginary status (e.g., perceptions, ideas, or concerns about the state of nature), or both. Thus, much political ecology is about recording environmental *destruction* and *construction*.

Part II

Conceptual and Methodological Challenges

In which efforts to measure the degradation of the environment are described, along with the difficulties and pitfalls that accompany such research, and wherein the "construction" of nature is discussed, along with its myriad meanings. Methodological caution and rigor are urged in measuring both environmental changes and imaginaries.

5

Destruction of Nature: Human Impact and Environmental Degradation

A walk into a German forest is an experience that for me raises contradictory feelings. Approaching down a dirt track into the thickness of the Schorfheide Forest (Figure 5.1) from the open farmed glacial landscape of the northern lowlands of Eberswalde, Brandenburg, the darkness is at first impressive. These trees are thick on the land, the sun blocked by the towering boughs up above. It is dark, quiet, and feels far removed from the orderly arrangements of the nearby poppy fields, canals, and autobahn. My first feeling is one of the weight of the growth, the darkness of primal nature. This is Germany's largest contiguous forest tract and the sense that I am "in the woods" is undeniable. This is a space to be preserved, protected, and fostered, if for no other reason than it provides respite from the expanding commercial landscapes of the world beyond. The forest, as it turns out, is a biosphere reserve, recognized by UNESCO as a site worthy of global attention and preservation.

Figure 5.1 Schorfheide Forest, Eberswalde, Germany

Coming to a halt in the shade and gazing into the thickness, however, I am immediately struck by the incredible creepy homogeneity of the scene. These trees are all of the same age, all equally thick and tall. They are almost all pine (*Pinus sylvestris*) with occasional specimens of European beech (*Fagus sylvatica*). Absent from the highly acidified forest floor is any significant undergrowth or any species other than the quick-growing, harvestable trees. Few gaps are open in the canopy and little disturbance or secondary succession is evident. There are no signs of natural predators, or faunal diversity of any sort. I am, in that instant, overwhelmed by the strange feeling that this forest represents a degraded scar, a deprived and empty landscape, where commercial and state interests have halted chaotic natural processes in their tracks. The area has been used as hunting grounds from the time of Friedrich Wilhelm IV and even the former East German leader, Erich Honecker, and his comrades bagged game in these woods. The forest in that moment conveys the feeling of a sterile and overworked cornfield, laid bare by the ravages of a rather narrow set of political and economic demands.

On walking a few paces off the dirt track into the woods, however, a third discovery is inevitable: the tremendous order that such an effort in forest farming entails. The cathedral-like architecture of the place is remarkable, with pillars of trees arranged in perfect rows. The sense of control, foresight, and care is carved into the orderly lanes of trees arranged with systematic breaks. This forest is actually ground zero for the development of modern commercial forestry techniques, and as early as the 1830s standardized techniques for growth and harvest were being implemented on this landscape, purging undesired species and enforcing even-age planting and harvesting. This is no forest at all, but the dream of an engineer, a social construction of what a forest *should* look like, made real by political planning on an extremely large scale.

And yet, walking further off the track into the trees, signs of uncontrollable chaos emerge from the neat patterns of the planner's mind. Openings in the woods do indeed appear, windfall gaps and depressions from ancient glacial movements. These are filled with a range of unintended species, which, as they push their way up to the light, begin to crowd out the even rows of commercial trees. Animal species *are* here, many of them indigenous to the region, many more migrants, and a great many feral descendants of human introductions and experiments gone wrong. Even the raccoons that run wild in the area are descendants of some that escaped from a fur farm in the 1940s – an uncontrollable natural experiment exerting its own non-human will.

This raises puzzling and immediate questions in my mind about what is natural and what is not, what should be preserved and what should not, what is degraded and what is not, what can be controlled and what cannot. Is the Schorfheide Forest a natural wilderness, to be preserved from the depredations of development that are beginning to sweep the now unified eastern part of Germany? Is it a degraded scar, which demands restoration and disconnection from the institutional mechanisms of utilitarian forestry? Is it a social construction, revealing the human imagination of nature enacted onto the land? Is it a chaotic tableau, in which non-human species produce unintended and normatively undecipherable outcomes, despite humanity's best efforts at control? The answer, of course, is that it is all of these.

But in investigating political ecological process and seeking explanations of environmental and social change, this answer ("well . . . it's complicated") is insufficient. If an argument is to be made that the region is a victim of utilitarian extraction that has caused environmental decline and loss of species, which to be sure it has, then degradation – *destruction* of nature – must be defined and methods of measurement must be devised. Has there been a decline in natural productivity, biodiversity, or usefulness? If one is to argue that the forest is the product of a specific normative imaginary (a vision of what a forest ought to look like), then the history of describing, categorizing, and organizing its environmental systems – *construction* of nature – must be demonstrated. Who controls the language and normative assumptions of how a forest ought to look? The burden of explanation and investigation changes, depending on the approach to environment and ecology that one selects.

In the great diversity of research, political ecological questions and answers have depended predominantly on measuring or revealing one or both of two processes: the *destruction* and *construction* of environmental systems or landscapes. In this chapter, I will briefly outline the ways in which political ecologists have evaluated "destruction" of nature, human impact on the environment, and land degradation. The following chapter assesses the ways in which the ecology is "constructed" by humanity through systems and categories of knowledge. These two views have been extremely useful and are prerequisite to any understanding of human–environment interaction.

There are, however, serious tensions between wanting to claim that normatively undesirable environmental outcomes are "unnatural" (e.g., land degradation is an avoidable and bad thing) while wanting to show that the way we view what is "natural" is predetermined in the first place by social and cultural concepts/filters/structures. I will later suggest, therefore (in Chapter 11), that these two approaches be supplanted by a synthetic notion of environmental "production," which takes seriously the normative implications of land degradation, recognizes the socially constructed character of the conceptual apparatus for understanding nature, and is sensitive to the natural system components that participate in socio-environmental change.

The Focus on Human Impact

Recognizing and understanding the destruction of natural systems is an integral part of political ecology. When starvation occurs in West Africa because of soil depletion and nitrogen deficiencies in production caused by peanut farming, for example (Chapter 4), the political ecological claim is that the productivity of the land has decreased from a previous state, and has been altered by human exploitation. The impact of global seed oil markets cannot be said to have affected local ecology if the soil quality has not changed, if its lost productivity arose strictly from non-human causes, or if pre-existing practices would have resulted in the same outcome. Implied in this question of destruction is also the assumption that changes in ecology are ones from which the system might not be expected to recover for a long period of time, or indeed ever.

In this sense, there is a great range of "destructions" that have seized the attention of political ecologists. Soil erosion, deforestation, desertification, biodiversity loss, water pollution, as well as atmospheric and climate changes are all common targets for research attention. These environmental crises are selected in part because they are of pressing concern to producers around the world who, in making a living off the land, encounter them daily and are placed at disproportionate risk. But perhaps more importantly, these topics are of central interest because they are important to *apolitical* ecologists, whose dominant narrative – that people destroy ecosystems out of ignorance, selfishness, and overpopulation – is the central target of political ecology's critique.

By responding to the environmental cries of Malthusians and technocrats, political ecologists have inherited their project: identifying and explaining forms of regional environmental destruction or degradation. In so doing, they have taken on a difficult task. There is no doubt that environmental crises are ongoing. Globally, forests have decreased by nearly 20 percent since 1700 (Richards 1990); soil degradation is evident in every known type of environment (Rozanov et al. 1990); people infected with water-borne diseases number in the billions (Schwarz et al. 1990). Even so, many regional changes occur in environments already experiencing climatic and geomorphologic transformation, like long-term aridification or mountain growth. Such trends may drive land cover change and soil erosion quite independently from farmer practice or logging. So too, some human impacts serve to increase rather than decrease landscape productivity or diversity. Thus environmental destruction, a crucial concept for political ecology, presents serious analytical challenges.

The Sahelian crisis provides an instructive example. Here is a case where a disaster of historic proportions involved the decline of environmental systems on a regional scale and social and political upheaval for countless millions of people. Popular accounts put the blame on nature and population, holding that scarcity of resources, spread too thinly during an inevitable crisis, led to mass starvation. Political ecological accounts, on the other hand, asserted not only that socio-economic and institutional changes had made poor communities and households more vulnerable to scarcity (holding less land and with fewer redistributive moral networks) but further that political economic change had made ecological systems more vulnerable to degradation. The progressive pressures placed by marginal communities on the land were environmentally destructive, causing declines that were difficult, if not impossible, to reverse in the middle term.

But this destruction occurred in a highly variable environment, itself subject to serious and frequent drought. What if soil loss would have happened anyway? How are human influences determined amidst uncertainty and variability?

Defining and Measuring Degradation

In as straightforward a definition as is available, Johnson and Lewis define land degradation as "the substantial decrease in either or both of an area's biological productivity or usefulness due to human interference" (Johnson and Lewis 1995, p. 2). This might include reduced catches from fisheries as a result of increased effluents from on-shore activities like farming and industry. It might include reduced production of crops per unit of land and labor as a result of decreased soil potential arising from over-cropping and reduced fallow. And it might include reduced grassy animal forage because of the plantation of unpalatable plantation species on a pasture.

Note that in many cases, a system's "degradation" may be a loss of one capacity in exchange for another. Blaikie and Brookfield point out:

> As a perceptual term, however, [land degradation] is open to multiple interpretations. To a hunter or herder, the replacement of forest by savanna with a greater capacity to carry ruminants would not be considered degradation. Nor would forest replacement by agricultural land be seen as degradation by a colonizing farmer . . . Since degradation is a perceptual term, it must be expected that there will be a number of definitions in any situation. (Blaikie and Brookfield 1987, p. 4)

In that sense, the selection of criteria is a specifically political choice, conditioned by the purpose of investigation and the categories of concern of the researcher. The range of possibilities is endless, but some categories of importance to political ecologists include land degradation as:

- loss of natural productivity
- loss of biodiversity
- loss of usefulness
- creating or shifting risk ecology

Each of these is measured differently and each can be evaluated in multiple ways. The pitfalls in such measures are several, however, and the degree to which ultimate and measurable "land degradation," free from political assumptions, can be established is debatable.

Loss of natural productivity

The productive potential of a fishery, a field, or a pasture is determined by a number of factors. In the case of terrestrial environments, direct loss of soil through erosion has historically been equated with loss of productivity, though this is not necessarily an effective proxy; highly eroded landscapes may remain productive, while low productivity can occur in areas where there has not been significant soil loss. Degraded environments are better understood as those showing decreasing quantities of important soil nutrients, increasing levels of salinity, and loss of surface biomass (Johnson and Lewis 1995). These conditions might be measured directly, through soil sampling and surface survey, or indirectly through measures of downstream siltation, remote sensing using air photos, or analysis of satellite imagery (Kumar et al. 1997).

To demonstrate a loss of productivity, however, especially as induced by human agency, requires more than a measure of current conditions. It further requires either a measure of conditions over time in one place, or a comparative spatial assessment under varying uses. In either case it is necessary to determine the underlying and baseline conditions (bedrock, rainfall, slope, etc.). For changing soil conditions, there is a wealth of available techniques, ranging from direct observation and local histories using standardized classification schemes (Ovuka 2000), to indirect examination of downstream siltation using stratigraphy and chemical analysis (Zhang et al. 1997).

Loss of natural productivity can also be measured through changing conditions in biotic land cover away from "ecological climax" conditions deemed to be the "natural" vegetative state. This approach proceeds from the traditionally accepted ecological notion that, under given climatic conditions (prevailing precipitation and temperature within a range of inter- and intra-annual variation), if left alone, land cover vegetation achieves a relatively stable and predictable state through the process of succession. Sandy semi-arid regions of India, for example, which are subject to grazing and cutting, tend over time to be recolonized by low-lying herbaceous species, later by fast-growing shrubs, and eventually by slow-growing "climax" *Acacia* species. Land degradation, it is theorized, is succession in reverse, where climax species are removed, leaving only faster-growing cover, closer to the ground (Kumar and Bhandari 1993).

Following this logic, evidence of the existence and extent of degradation can be gleaned by examining current biotic structure on the ground. Measures of degradation using this method can employ direct floral surveys at sampled locations or work through remote sensing platforms to determine general patterns of vegetation cover. These are commonly supplemented with measures of overall surface biomass, again

Table 5.1 Land cover changes (in square kilometers) in the southern Yucatan peninsular region

Land cover	*1969*	*1987*	*1997*
Forest	11,042	10,356	10,068
Secondary growth	111	634	845
Agriculture	228	391	468

Source: Turner et al. (2001, p. 364)

either through direct measuring of sample areas, or through the use of satellite imagery to create land cover images and biomass density maps, using the Normalized Difference Vegetation Index (NDVI) – a ratio of spectral reflectance denoting biotic production. These surveys and images can be used to quantitatively compare land cover over time, suggesting trends in the productive potential of the landscape (Eastman et al. 1991). This can be further supplemented with oral history, written records, and other supporting documentation.

Such approaches commonly reveal the complexity of trends in land degradation and recovery. For example, Turner et al. (2001) have measured land cover change in southern Yucatan, an area where protection of forest for ecotourism development is sometimes contradicted by the construction of infrastructure designed to support that tourism. Pressures on the forest, like road development and farming, are tied to increased integration with global tourism and commodity markets. But has that development resulted in overall degradation and decline of native forest? Here, evidence of land cover change is obtained through remote sensing and local history to present a complex picture. While forest cover is shown to have declined from 1969 to 1997 by around 9 percent and agricultural land nearly to have doubled, the coverage of secondary growth – areas in regrowth towards mature forest – expanded almost eightfold (Table 5.1). This suggests that human disturbance of forest has increased over the last three decades, with areas under "degraded" or secondary growth expanding. Even so, this in no way necessarily represents a permanent trend, since forest is returning. More details are required – what species are in decline and what is the rate of forest recovery in secondary growth? – before a comprehensive picture of environmental change is clear.

Loss of biodiversity

As an imperative in global conservation, species diversity is of increasing concern. For resource-dependent people also, the diversity of the landscape may be of crucial importance, since the range of available species on which people depend can be far more important than the soil structure or overall biomass of an area. Biodiversity loss also provides a window into the potential long-term effects of human impact. Biodiverse and heterogeneous systems are complex and can potentially withstand and recover from intense human and environmental shocks. A decline of diversity may be the leading edge of serious and sustained declines in later productivity.

Measuring biodiversity on the ground is a difficult and time-consuming exercise. It requires careful sampling of the landscape to establish a representative set of plots in which to work. Some techniques require researchers to scour hundreds of 20 meter plots over several hectares (Dallmeier 1992), while others demand the survey of thin swaths of land along enormous transects (Gentry 1986). In either case, work must carefully document the number and richness of species, with specific attention to important indicator species that are most vulnerable to disturbance. In a singularly heroic example, the Idaho National Engineering and Environmental Laboratory created a reserve to explore species cover, density, and frequency over time. Examining 79 permanent plots over 45 years, they were able to determine relationships of disturbance, recovery, invasion, and diversity (Anderson and Inouye 2001).

Little of this sort of long-term data exists for political ecological analysis, precisely because its research questions are directed at *in situ* (rather than experimental) socio-environmental conditions. But sustained and field-based biodiversity assessment remains an important part of political ecology, especially when conducted in collaboration with indigenous communities. Rocheleau's research in the Dominican Republic, for example, has revealed diversity change resulting from economic and institutional change in remarkable detail. There, the introduction of a fast-growing cash timber species (*Acacia mangium*), when linked to economic development initiatives, was demonstrated to have transformed biotic assemblages not only in the forest, but in producers' fields, pastures, and gardens (Rocheleau et al. 2001).

Loss of usefulness

Assessing whether or not an environment is more or less useful as a result of human action is in many ways the most direct, practical, explicit, and politically honest approach to measuring environmental destruction. When a pasture cannot be used for its traditional purpose of grazing, a field for growing crops, and a forest for providing socially and economically important tree species, some kind of important change has occurred.

Measurement of usefulness is, however, not altogether straightforward. Is land more useful when it is providing the highest return or providing the greatest collective benefit to a community? Is it achieving highest current return or lowering risk of future disaster? Is it measured in financial return or by some other criteria?

Determination of an area's "appropriate" use is also explicitly political. As noted above, turning forest into pasture or vice versa may be seen as degradation or improvement depending on the community and its resource use priorities. Despite this, or perhaps because of it, the "use" approach is perhaps the most pervasive one in formal management policy. The Clean Water Act in the United States, as a leading example, requires biannual assessment of American waterways to determine whether or not they are meeting designated use criteria (fishable, swimmable, etc.), thereby codifying explicit social goals into environmental management (Adler et al. 1993).

Methodologically, assessment of changing "usefulness" is also perhaps the most viable approach, since even where explicit and detailed soil histories, land cover descriptions, and diversity profiles are unavailable for past landscapes, land uses are commonly recorded. Oral and written histories, photographs, and management

Box 5.1 Ecology matters in Rocheleau's Dominican Republic

Any paper that begins by stating that "forests . . . are inscribed with social relations," suggests a research effort into discourses, gender relations, struggles between states and localities, popular organization, and non-governmental organizations. And to be sure, in two key publications (Rocheleau and Ross 1995; Rocheleau et al. 2001) Dianne Rocheleau and her several colleagues have assembled compelling accounts of conflicts over control and access to resources in the Dominican Republic. For example, they show that the farmers of the region have acquired (or recaptured) their land in a series of non-violent struggles using civil disobedience, and that now, allied into rural federations, they continue to struggle for land rights.

But what sets this work apart is its sensitivity to the way ecological conditions matter in explaining the way things turn out. The research shows that while the adoption and impact of *Acacia mangium* – a highly politicized, fast-growing plantation species introduced to provide commercial opportunities for poor smallholders – is affected by social factors, it is also determined by local landscape ecology. This means that while land tenure, gendered household division of labor, and affiliation with political/development organizations are important to understanding the rate and trajectory of tree cover change, so are things like existing plant biodiversity and species composition on farms. Rocheleau's investigation, which began as a short-term exploration and turned into a multi-year survey, details the way supposedly deforested areas, targeted for monocultural afforestation, are actually diverse and species-rich.

As Rocheleau explained in 2003, "Rather than being content with this profusion of social data, I found myself absolutely *needing to know* about the biodiversity of these patches and ribbons of forest in a regional landscape quilted by the Federation members into a distinctive socio-ecological formation, a regional agroforest rooted in community, a shared history of struggle and visions of a possible agrarian future."

These concerns reflect Rocheleau's long-term commitment to revealing the intertwining of human and non-human ecologies. Indeed, this work makes it difficult to distinguish the independent influence of one separate from the other. The simplification of ecology (e.g., biodiversity decline in house plots and fields) and the marginalization of women (e.g., increased labor burden and less control over resources) are tied together. Social movements and development strategies that influence one, influence the other. Rocheleau reminds us that just because forests are inscribed with social processes, they are not suddenly deprived of biophysical ones. Instead, this work asks us to consider how illusory the boundaries between these are.

Rocheleau regrets only that publishing this kind of integrated analysis is difficult, since scientific conservation biology and political ecology have too little understanding of or patience for one other and too rarely communicate. As she explains, "the two banks of the river are treated as separate continents."

records can all provide some kind of historical picture of the changing useful capacity of environments. More standardized data comes in the form of crop yields, stocking rates, and economic value. These kinds of data can be deceptive, insofar as they vary in response to a range of forces beyond ecological conditions, but they do provide a starting place for assessing environmental change.

The anthropocentrism of the approach is worth noting also, however, since ecologically impoverished landscapes, lacking in diversity, and providing few ecosystem services, may well be serving important uses, if only as sinks for waste or provision of a small range of resources. This notion, that an area might act as a "sacrifice zone" for other areas, is an important one in land degradation study because it indicates the complex issue of *creative destruction* – where some uses and functions are lost to benefit others (Johnson and Lewis 1995).

Socio-environmental destruction: creating or shifting risk ecology

In many cases, the research question is not whether land use or management has altered productivity, diversity, or usefulness, but instead whether it has led to an increasing vulnerability of an area to destruction (fire, erosion, mass slumping) or created new risks or hazards for local residents. This is of particular interest in political ecology if the level or location of risk is shifted onto vulnerable or disempowered populations.

If, for example, changing land prices drive poor people – with relatively little latitude of choice for house construction – into building houses on steep hillsides, immediate erosion may not be visible or evident, but the risk of catastrophe during an abnormal rain event is definitely increased (Smyth and Royle 2000). In agrarian environments this process might take the form of increased extraction from landscapes that, though it shows little immediate vegetative cover change, may be dramatically affected by a major climate event like a drought.

In many settings, changes in the ecology of production may increase productivity but also create increased risks for cultivators in terms of health and welfare (Shiva 1991), as is the case where farmworker exposure to toxic chemicals is precipitated by changing agricultural practices (Pulido 1996). Much recent research in the area of environmental justice, which focuses largely on the location of man-made hazards like industrial plants or waste sites in the proximity of disempowered populations, draws attention to ecologies where risk is spatially "externalized" from one group to another (Szasz 1994; Cutter 1995; Miller et al. 1996; Been and Gupta 1997; Pastor et al. 2001). Thus, the production of risks and hazards in the environment represents a form of environmental destruction, where normal risk situations are made acute or shifted to specific people or groups through ecological change.

Limits of Land Degradation: Variability, Disturbance, and Recovery

Despite their common employment, these methods of degradation assessment have flaws and can be extremely misleading. Whether for measuring loss of productivity, diversity, or usefulness, new understandings of ecological dynamics raise questions about degradation as a meaningful approach to understanding human impact and disturbance.

What baseline? Non-human disturbance and variability of ecological systems

Many biotic systems are given to tremendous variation both within and between years, and most natural systems, even when isolated from human influence, are highly dynamic. This is especially true in the tropics, where important wet–dry cycles and frequent atmospheric anomalies mean years with luxuriant growth often interrupted by long cycles with little growth or development. Some vegetative cover trends are long-term responses to regional climate change and may in no way reflect human impacts over time.

Consider the Boundary Waters Canoe Area of Minnesota and Ontario. Currently an area of thick marshes and forest, the region shows tremendous variability in its long history since it was locked in tundra 10,000 years ago. As ecologist Daniel Botkin describes, lakebed pollen records suggest that following the last glaciation:

> the tundra was replaced by a forest of spruce, species that are now found in the boreal forests of the north, where they dominate many areas of Alaska and Ontario. About 9,200 years ago the spruce forest was replaced by a forest of jack pine and red pine, trees characteristic of warmer and drier areas. Paper birch and alder immigrated into this forest about 8,300 years ago; white pine arrived about 7,000 years ago, and then there was a return to spruce, jack pine, and white pine, suggesting a cooling of the climate. Thus, every thousand years a substantial change occurred in the vegetation of the forest, reflecting in part changes in the climate and the arrival of species that had been driven south during the ice age and were slowly returning. Which of these forests represented the natural state? (Botkin 1990, pp. 58–9)

In more recent history, the American tropics and subtropics underwent climatic changes during the pre-Columbian period, with drying and wetting trends over several thousand years. These were further punctuated by inter-annual variations and spikes, probably linked through teleconnections – pressure and temperature interconnections around the globe – to South Pacific warming and cooling cycles we today know as El Nino (Lentz 2000). In this context, it is particularly difficult to assign environmental impacts to humans, either indigenes or colonial invaders, in any simple way.

Even on shorter time scales, production, disturbance, and regrowth may cycle repeatedly, meaning that current conditions, even where there is no influence or impact from people, may reflect a transitional state. Fire cycles in the US, for example, show 25–30 year recurrence in mixed conifer woodlands (Agee 1993). At any moment a regional forest may be in a successional state far from climax.

The implications for political ecology are evident. Current environmental conditions are merely a snapshot of complex change. Determining whether apparently low levels of current production represent a significant trend, or even a product of human action, requires careful attention to temporal variation and the establishment of meaningful baselines for comparison.

Landscapes are also ecologically and spatially heterogeneous, or "patchy." A relatively small area of forest may have spaces dominated by a few species, bare areas, and patches of diversity. Fisheries are marked by similar submarine diversity, with "sweet spots" and breeding areas interspersed throughout open ocean (St Martin 2001).

This tendency towards environmental heterogeneity also holds implications for political ecology. Claims of degradation or environmental change are inherently *scale-specific*. Pre-existing landscape diversity can be seriously decreased by large-scale transformations, as where highly varied Brazilian tropical forests have been converted to soybean production on a large scale. On the other hand, some disturbance may enhance ecological diversity depending on the scale of analysis. An area of forest may have experienced serious decline in floral biodiversity in the wake of human cutting or burning, for example, with new herbaceous species invading the once-shaded forest floor. The adjacent spot, however, may be covered with indigenous canopy. Together, the two spaces actually represent greater diversity than if the patch had not been burned. Some woodland areas, under heavy but spatially scattered human usage, have been found to be particularly species-rich (Blumler 1998).

What impact? Variable response to disturbance

Equally important, many biotic systems actually depend on disturbance, including fires, windstorms, or herbivory, for the maintenance of important species and the development of what were historically considered pristine "climax" conditions. Forests of the North American Pacific Northwest, for example, are adapted to the periodic fires that sweep the region such that many important species depend on fire to clear out competition, allow cones to open, and seeds to germinate (Agee 1993). Considerable work has also shown that anthropogenic disturbances can mimic 'natural' disturbances. This means that human disturbance, including cutting, grazing, and digging, may increase biotic productivity and maintain climax conditions. Reading non-climax conditions as evidence of human impact, therefore, is to make an oversimple assumption in many ecosystems.

As with biotic productivity, ecologists have long recognized that biodiversity is also enhanced by periodic, non-catastrophic disturbances (Huston 1979; Petraitis et al. 1989). As long as the interval between disturbances is less than or equal to the recovery time, transient communities, which typically comprise more species per area than non-disturbed communities, will dominate a given area. Evidence from semi-tropical and tropical woodlands and savannas, for example, has shown that even under conditions of continuous grazing and browsing by livestock, biodiversity can be maintained if not enhanced (Huston 1979; Turner 1998; Oba et al. 2000; Fuhlendorf and Engle 2001).

Differing types and intensities of disturbance may produce differing results. Where heavy grazing by cattle may have an impact on the succession of grasses and herbaceous species, for example, goat and sheep browsing might affect tree sapling development. Intensive plantation agriculture differs from long-fallow rainforest farming in its dampening of diversity. In industrial contexts, the conversion of wetland to agriculture is considerably different than its conversion to pavement, especially for faunal species making use of the ecosystem.

Box 5.2 Colonial complexity in Crosby's *Ecological Imperialism*

Eight of the ten most common lawn and golf course grasses in the United States, which make up as much as a quarter of all urban land cover, are not indigenous to North America (Robbins and Sharp 2003). Indeed, these most "American" of all landscapes are actually quite foreign. Tens of thousands of other plant species, which make up much of the daily landscape of Canada, Mexico, and the US, marched into the environment long ago, along with their allies, human beings. That there is a relationship between this pattern of environmental change and the pattern of colonization and imperial control of the New World in the "age of exploration" is the central thesis of Alfred Crosby's landmark book, *Ecological Imperialism: The Biological Expansion of Europe, 900–1900* (Crosby 1986).

In this clearly written, compelling, and well-researched environmental history, Crosby argues that Europeans brought with them a set of *portmanteau* biota, including diseases, songbirds, housecats, weeds, cattle, and horses, which advanced in a mutually supporting phalanx across the "neo-Europes" of the Americas and Australia, displacing native species and supporting the coercive efforts of human occupation.

Most of this transformation and accompanying mass extinction occurred in the colonial era, Crosby explains, since previous efforts at invasion (in the crusades and the conquests of Asia, for example) entered environments where existing patterns of disease, flora, and fauna, were well enough integrated and networked to provide a solid defense. The neo-Europes were ecologically more vulnerable to invasion.

It might seem a "natural" jump to extend this explanation to account for human success in the New World and to argue, in other words, that Europeans came to dominate the world *because* their ecologies allowed it and indeed encouraged it. Such arguments have been made elsewhere (Diamond 1997). Crosby intends no such thing, however, and draws a rather subtler historical conclusion.

He argues that environments and people are mutually produced, leading to complex strengths and vulnerabilities of ecological systems. The simplification of New World ecologies, for example, which allowed the invasion of Old World species in places like the South American pampas grasslands, had been faciliated by the hunting and landscape modification (e.g., fire) of precolonial native people. Thus, "advanced" whites did not ecologically triumph over "inferior" indigenous human ecologies. Crosby's thesis, in this way: "places the Amerindians, Aborigenes, and Maori, on the one hand, and the European invaders, on the other, in a fresh and intellectually provocative relationship: not simply as adversaries, with the indigenes passive, the whites active, but as two waves of invaders of the same species, the first acting as shock troops, clearing the way for the second wave" (Crosby 1986, p. 280).

Such a theory, with cultural ecological evidence of native landscape influences growing (Doolittle 2000; Denevan 2001), is a refreshing step away from notions of European cultural and environmental superiority. As such, the book represents an elegant and ecologically solid example of political ecology.

The capacity of differing systems to absorb human disturbance also varies greatly. Consider a rainforest cleared in Nicaragua with chainsaws and bulldozers for timber. The apparent visual effect is dramatic. The landscape of the forest is covered in fallen and dead tree and leaf litter. The sky, usually roofed by thick layers of canopy, is open to the punishing heat of the tropical sun. Even so, the overall diversity structure of the forest is little affected and, when it recovers, will retain the system components of its pre-disturbance state. Plantation of the region into bananas, however, may over time create significant structural changes in local diversity, such that, when it is abandoned, the forest will return, if it returns at all, to a considerably different state (Vandermeer and Perfecto 1995).

Different systems, therefore, may respond very differently to human impacts, depending on the ecosytem characteristics, including initial diversity, climate, annual and inter-annual variability, and the disturbance profile. Some systems are highly stable; their productivity (or diversity or whatever other indicator of destruction/degradation is of concern) may decline slowly with impact, while others may be sensitive to low levels of impact. Such sensitive or "fragile" systems often exist where productivity of the system is low, as in arid ecosystems and grasslands. Still other systems reflect more complex dynamics, and are able to maintain moderate levels of impact or extraction with little or no effect upon productivity until a threshold is passed, when such systems may change rapidly. Current ecological research suggests that many natural ecosystems behave in this fashion, not showing signs of degradation until rapid change is seen.

The implications for researchers interested in political and economic influences on environmental change are several. First, such dynamics suggest that certain intensities of human uses, local peasant extraction or forest harvesting, for example, may have little immediate or sustained impact on an ecosystem, while increases in intensity of use or extraction, due to falling commodity prices or failed common property management rules, may have sudden and precipitous effects (Reynolds and Stafford Smith 2002).

It is also likely that different kinds of extraction/use on the same landscape follow different impact curves from one another. Heavy or sustained grazing in a forest may result in reduced productivity as animal numbers increase until it reaches a high-intensity final saturation point. The impact of tree-cutters on the same forest might be considerably different, however, as an increasing number of extractors have little overall impact on the forest until a "breaking point" is passed and forest reproduction is significantly reduced. Socio-economic changes may create demands for new resources from the same ecosystem or deliver new forms of waste. The change in type of impact may be more important than the change in intensity.

Significantly, a system can demonstrate tremendous variability but still remain highly resilient. Imagine a productive pastureland that experiences short-term acute heavy grazing by a passing herd of cattle. The biodiversity, productivity, and usefulness of the system might all decline precipitously. The landscape may recover quickly and fully, however, with the arrival of the first rain, especially where ungerminated annual seeds and deep perennial rootstocks lie dormant below the surface.

The implication of this variability is crucial for political ecology. It suggests that systems can withstand and thrive under a range of human uses, especially those of

Figure 5.2 System resilience in coppice recovery of savanna trees. This khejri (*Prosopis cineraria*) tree is harvested heavily during the dry season (left) but recovers after only the first rainfall (right)

moderate intensity, under which the ecosystem may have evolved and to which it may be adapted. This means that the traditional subsistence livelihoods of the world's poorest people, including smallholders, slash-and-burn agriculturists, and nomadic herders, may have no serious long-term negative impacts on ecosystem productivity. Rapid changes in such systems, on the other hand, including increases in energy throughputs, higher levels of extraction, or new species, may have serious and sudden deleterious results. Thus, while ecological research suggests caution in attributing and determining degradation, it can support the general political ecological principle that subsistence communities are not a threat to ecosystem sustainability until larger developments and socio-economic changes alter key elements in their use of the landscape.

Can we go back? Variable recovery from disturbance

The impact of a human disturbance, though it may lead to decreased productivity, diversity, or usefulness in the short term, may not represent a sustained, permanent, or seriously irreversible impact. Different systems may recover from disturbance in a range of ways. Some demonstrate temporary decreases in productivity, followed by speedy recoveries. Others remain low in productivity for long periods. Still others recover, but in an altogether new state, with a different mix of species, never returning to their former state. Experimental research reveals, for example, that relatively rapid recovery of ecosystems can follow the removal of disturbance pressure, but that dystrophic (highly leached and low-nutrient) soils tend to seriously retard recovery time (Harrison and Shackleton 1999); many systems are remarkably resilient, but rates of recovery are determined by complex edaphic conditions.

For political ecology, this serves as a cautionary lesson; not all environmental destruction is permanent. Even so, some ecosystems are extremely vulnerable to long-term transformation, and some of what determines the recovery of ecosystems inheres in the environmental conditions of soil, moisture, diversity, etc.

Many ecosystems, however, exhibit further characteristics that make their dynamics especially complex: modality and hysteresis (Lockwood and Lockwood 1993). Modality is the existence of multiple distinct states that a system can encompass. Hysteresis is a condition where processes of degradation are not reversible simply by eliminating disturbance, which may instead lead to new states.

A grassland, for example, may have a desirable "climax" condition, with rich and diverse coverage, dominated by perennial grass species, which is maintained under herbivory and which is relatively productive. The ecosystem may exhibit *modality*, however; it may have a second equilibrium state, where annual grasses dominate, with lower overall diversity and productivity. The grassland may further exhibit *hysteresis*, where the recovery from disturbance is not always reversible by simply stopping the disturbance event. Indeed, recovery from disturbance may lead to an altogether new state (Westoby et al. 1989). Under traditional management of such a grassland, where heavy grazing causes a decrease in productivity, achieving a return to the original state is usually thought to only be a matter of removing the grazers from the land and waiting for the perennials to return. Since the system has multiple states, however, and the path back from a disturbance is often different than the path forwards, this may not be the case. After some rest, the pasture may not revert back to perennial grasses but may instead become dominated by other woody species and annual herbs. In this sense, some impacts may be considered "irreversible" within reasonable human time scales.

On the other hand, it also means that human activity can produce new landscapes through management, and intimate knowledge of ecosystem transition, the kind of knowledge that many local producers around the world have, has allowed ecosystem "engineering" for millennia. Re-engineering of these landscapes is possible, especially when social, political, and economic stressors drive changes in the type and intensity of use.

Consider the coastal grasslands of the Eastern Cape Province of South Africa, a crucial economic resource that provides grazing, building materials, and medicines, and is central to the lives of rural people throughout the region. The ecology of the region is, moreover, incredibly diverse; among others there are high-quality grazing lands dominated by *Themeda triandra*, unproductive grasslands dominated by *Aristida junciformis*, and valuable collection lands dominated by *Cymbopogon validus*. Researching the origins of these landscapes, Thembela Kepe and Ian Scoones (1999) consulted local producers and reconstructed the history of environmental change, both in terms of the biotic communities on the land as well as the institutions that produced them. Their work reveals that these varying ecologies

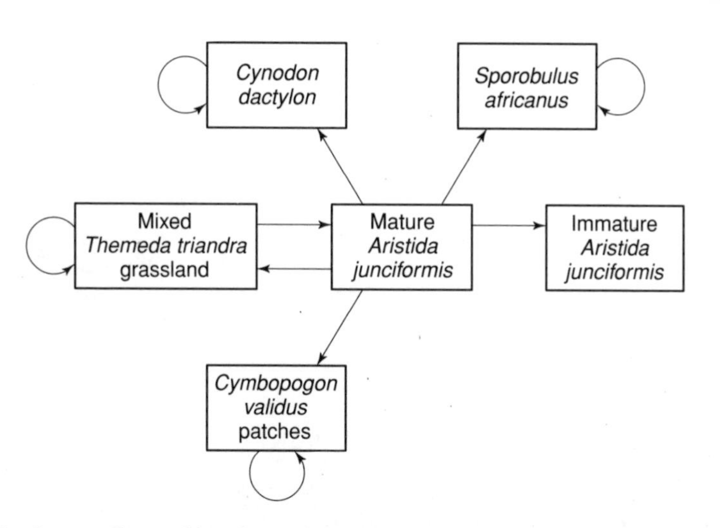

Figure 5.3 State and transition for ecosystems in the Mkambati area (Reproduced from Kepe and Scoones 1999, figure 5)

were all products of transitions from other environmental states, encouraged or discouraged by locals through burning, grazing, and enriching or disturbing soils (Figures 5.3 and 5.4).

Aristida junciformis grassland, for example, may be transformed through seasonal burning and periodic rest into highly productive *Themeda triandra* pasturage or into *Cynodon dactylon* grasslands by soil enrichment. *Themeda* pasture, on the other hand, can return to a state of *Aristida* domination (lower productivity and usefulness) through heavy grazing, whereas heavy grazing sustains *Cynodon* grasses. The ecology of landscape change is complex but amenable to careful research.

As Kepe and Scoones go on to show, moreover, the transitions between various states are achieved by different groups, with different goals, all working to shape the grassland in different ways. The resulting diversity of outcomes does indeed include instances of difficult-to-reverse destruction including transition to low-productivity, immature grassland dominated by the marginal species. But so too, many other transitions are possible, including to productive perennial pasture, literally produced by human action. A political ecology of human impact that takes seriously the complexity of degradation and recovery dynamics can, therefore, point not only to the political drivers of degradation, but also to the political possibilities of sustainable management.

Methodological Imperatives in Political Analysis of Environmental Destruction

Thus, to paraphrase Marx, people make landscapes, but not always those of their own choosing. Systems are driven to new states, some recover, others don't; some take new forms, which in turn enter new states, or return to earlier mixtures of elements. In examining environmental change we should perhaps think in terms of

Figure 5.4 Complex patchworks of grassland, forest, and scrub in Mpumulunga (above) are produced through traditional practices of burning for succession control (below) (Courtesy of Thembela Kepe and Ian Scoones)

a political ecology of *transitions*, rather than unhealthy/healthy or broken/fixed ecosystems. It is probably better to think of a political ecology of *production*, rather than of destruction, therefore (Chapter 11).

However, even while the measurement of degradation is a complex matter, as is the burden of its proof, the concept is essential to explicitly normative and political approaches to the environment. Fertile and productive land can be rendered nearly unusable. Diverse forests can be transformed into monocultural plains. Plant and animal species can be eliminated. Indeed, they are disappearing at rates unprecedented in human history. These changes are serious and sometimes extremely difficult to reverse. Research that engages crucial questions about the interaction of human and non-human processes can ill-afford to ignore such trends, despite the complexity of addressing and interpreting them. Such trends can be measured, moreover, at least in local and regional contexts, and the reward for careful and thoughtful research in this area is a more open door onto sustainability in the future.

In sum, research in regional political ecology, especially investigation into system changes that attribute environmental transformation to social and political forces, requires an acute attention to the ecological characteristics of the landscape in question. Researchers must:

- establish the overall type, rate, and direction of, possibly multiple, environmental changes
- identify the drivers of that change, human and non-human
- determine the environmental context in which such changes occur, including pre-existing variability and dynamics
- explore the specific impacts of various practices in terms of their intended and unintended effects
- examine the capacity, rate, and direction of routes of ecological recovery following changes or cessation of impacts

In truth, political ecology has not always been entirely attentive to these methodological or conceptual problems. This reflects less on the approach than it does on the enormity of the undertaking. The political ecologist's goal, to show the influence of political economy on such already complex systems, is therefore absolutely realizable, but only with attention to ecological dynamics.

Returning, then, to the Schorfheide Forest, some specific, researchable questions *can* be asked. Is this forest more or less diverse than the landscape that stood in its place three centuries before? Are areas within the forest that go unplanted or unburned more or less productive or diverse than those that are? What policies have led to decreased or increased productivity and diversity? What political interests and power coalitions produced such policies, and what user communities were removed either from forest, decision-making, or both, in the process? What, in sum, is the relationship between power and environmental change? Rigorous and careful landscape assessment, perhaps using remote sensing, ground-level survey, and oral and written history, taking careful note of ongoing non-human influences and trajectories of change, can provide a political ecology of the forest.

But to have asked and answered these questions is not to have pursued the full range of processes and relationships at work. Bear in mind that the thick, ordered monoculture of the forest, despite its relative lack of diversity and ecosystem function, is one that many foresters, officials, and regional residents would historically have considered as natural, or at least as forest. This is by no means uncontroversial. Many environmentalists might argue that the monoculture farm forestry of the area is barely a forest, and by no means natural. So too, the definition of this kind of conservation landscape as forest reflects an intentional, if not conscious, social program that is an important part of scientific forestry history in Germany. The question one might ask, therefore, is: How did this specific notion of nature, this image of the forest, become the taken-for-granted one? How was the forest constructed and by whom?

6

Construction of Nature: Environmental Knowledges and Imaginaries

On a day several months into the Indian dry season, a forester stands in a low alluvial plain, looking out at the stones in the streambed and to the hills beyond. Pointing at a thick area of thorny trees on the opposite bank, he explains to me that the forest is returning in this area after years of abuse and neglect. Deep-rooted hardy tree species are securing the embankments and restoring greenery to the desert. The species responsible for this remarkable turn-around in the region are largely imported through global initiatives in scientific forestry. The tree, *Prosopis juliflora*, he explains, is salt-tolerant, nitrogen fixing, drought-resistant, and very productive.

Along with several other introduced species – *juliflora* was brought from the Mexican/US southwest a century ago – the tree has helped to triple forest productivity in the last 30 years.

For the forester, this increase in forest area represents not only an institutional victory, but also a personal triumph. Foresters in the crowded middle and lower ranks of the bureaucracy sometimes go decades without a promotion, watching projects develop with little or no success. To the degree that they are repaid for their effort, small bribes from local people and the occasional small "feast" from a local landlord are more common than official reward. The achievement of significant forest cover – the official goal of not only the Indian state but also of World Bank donors – is an important success for the local bureaucracy, one that will minimally assure the flow of already limited funding and support. In listening to the forester's story, it seems that, far more abstractly, this achievement of forest cover is a deeply internal and aesthetic pleasure. Greenery is good.

But standing again in the same streambed a few months later with a local herder, a member of the *raika* caste, an extended kin network of herders and livestock breeders who have lived in the region for countless generations, I learn that this tree cover represented something else entirely. The old man, leaning on a tall gnarled staff and shaking his head at the cover of *juliflora* across the rocks, explained that the tree was a hazard and a blemish, that it had no value and that it crowded out valuable grasses and forage. More to the point, he insisted that the trees represented no kind of forest at all; on the contrary, they had created *banjar*, degraded wasteland. This wasn't *junglat* (forest), this was simply *angrezi*, foreign *English* landscape.

Standing there amidst the mesquite, I experience exactly the same feeling of intriguing confusion that I do when walking in the Schorfheide Forest. There is simply no objective way to determine whether the trees at which I was gazing were forest or not. Forest, put simply, is not a *natural* phenomenon, object, or idea, it is a *social* one, forged by convention and context, and enforced by its very taken-for-grantedness. This becomes especially political when one considers that, depending on whether this bunch of trees is considered "forest" or "degradation," significant state and international resources will be invested in its protection or its eradication.

Such a realization, that an evidently natural object, idea, or process is, at bottom, an expression of the human imagination, suffused with political and cultural influences, is one that is fundamental to much explanation in political ecology. Examining historical and contemporary environmental discourse and environmental science, political ecologists commonly argue the environment we take for granted is actually *constructed*.

Why Bother to Argue that Nature (or Forests or Land Degradation . . .) is Constructed?

This approach is by no means a novel one. In *The Critique of Pure Reason*, nineteenth-century philosopher Immanuel Kant proposes a metaphysics where philosophical knowledge comes *prior to* experience. Radically, he suggests that our ideas do not conform to the objects of the world around us but that, rather, objects are constituted by the world of our ideas (Kant 1882). In the century since its publication, a wide range of philosophers and historians of science and knowledge have pursued the question, with a recent explosion of claims to the constructedness of just about everything.

Michel Foucault (see Chapter 2) led the most recent generation of critical scholarship pursuing this line of argument. By doing intense historical study of many taken-for-granted ideas, including insanity and sickness, Foucault shows that many concepts that we currently assume to be universally true simply didn't exist in other times and places. He demonstrates, moreover, that the promulgation of these novel categories of reality has consistently been linked with the emergence of new authorities and institutions empowered to manage, rule, care for, or otherwise control social life, including medical and penal systems made possible only by social invention of madmen and deviants. So systems of new knowledge necessitate new forms of social power and vice versa. Ideas are not powerful because they are true, Foucault insists, they are true because of power. This development marked a change in critical politics. Rather than simply pursue the goal of the 1970s bumper sticker, "Question Authority," Foucault and his followers pressed us to more radically "Question Reality," as the more recent bumper sticker invites.

The implications for environmental management are important because they direct our attention to the social origins of environmental processes and objects. Soil erosion, for example, is not a universal truth. Rather it is a social construction, invented in the historical moment when colonial land management authorities, state environmental bureaucracies, and other ecological elites were given the power to control other people's behaviors and property in the name of "soil conservation." Resistance against the imposition of such colonial controls, as was common in colonial Africa, was viewed by officials as environmental irrationalism by an ecologically destructive and ignorant native populace (Grove 1990). Soil erosion was a social construction that helped to secure colonial power.

This sort of investigation is extremely common in contemporary political ecology, but specification of the meaning of terms is often neglected. What do political ecologists really mean when they say something is "constructed"? Following philosopher Ian Hacking, this claim means that some kinds of environmental processes, concepts, ideas, or entities are not natural or inevitable, even if they appear that way, and the history of these phenomena can be traced, and their invention discovered, through analysis. Moreover, as normative researchers, political ecologists generally pursue this claim because they believe that these processes, concepts, ideas, or entities, in the current socio-political context, are doing pernicious work or helping to secure the power of an elite community. Moreover, the politics that govern the fate of natural systems are secured without resistance to the degree that this constructedness is hidden from view. Political ecologists suggest, therefore, that because this stuff (processes, concepts, ideas, or entities) is not inevitable and has history, it can be unmasked for what it is, reinvented, and changed for a better and more sustainable future. In any case, in political ecology, things are rarely what they appear.

Choosing targets for political ecological constructivism

Despite occasional pejorative writing, which characterizes constructivist accounts as dangerously relativistic efforts to disallow any search for practical solutions or any distinction between truth and falsehood, most constructivism in political ecology does not seek a dethroning of all that is real.

First, political ecological constructivism tends to focus specifically on those things that are most taken for granted. Favorite targets are concepts that slip into our thought worlds over time, cementing and directing how we categorize environmental reality without thinking about it: living metaphors that have become dead (Barnes and Duncan 1992). Through research, called "archaeology" in the terminology of Foucault (Foucault 1971) or "genealogy" in the language of Nietzsche (Nietzsche 1967), critics can create a history for things that appear to be without history, and so make them uncontrovertibly true. Soil erosion in the Himalayas, for example, makes a likely target for political ecological constructivism because it is an idea that is so universal and commands so much institutional attention, global funding, and academic research that it has become apparently true. Despite rigorous scientific research that contradicts the idea of a generalized and accelerated anthropogenic degradation of the region, the idea is still fundamental to regional development initiatives (Blaikie 1999). Discovering how, when, and why that is so, is an important research task, therefore.

Second, this kind of research tends to center on concepts and objects that, by being constructed one way, disallow alternative interpretations, and so mask political motivations and activities. Consider, for example, the "environmental refugee," a person not known to exist until recently, but now numbering in the millions, at least according to Red Cross documents, World Bank plans, and United Nations statements. As Saunders (2000) explained in an analysis of the historical genesis of the concept, this commonsense category, suggesting a person fleeing from environmental disaster and creating a nuisance in their destination location, is actually quite newly invented, though with some deep roots in earlier Malthusian concepts. More importantly, the concept also conceals the geopolitical conditions and machinations (those that political ecological chains of explanation seek to reveal) that set the conditions for the flight of refugees, turning a manifestly political process into an apparently (and deceptively) apolitical one. The importance of this sort of critique is not just that it denounces or refutes something that is "false," but that it also examines the processes and strategies by real people and agencies that go into making that thing "true." Such an unmasking hopefully makes it harder for such shenanigans to work in the future.

Third, political ecologists tend to choose as the target of their archaeology those constructions that contribute to constructing the objects of the world. Returning to the Schorfheide Forest, it is not simply the idea of "forest" that is being constructed, after all, but the actual physical environment of trees and ground cover. The imagined forest becomes the real one, and vice versa, through the enforcement of such constructs by powerful people over time. In this way, the line between objects and ideas is blurred.

Debates and motivations

The constructivist approach to the environment is politically and intellectually valuable, but it is not uncontroversial. To say that a phenomenon like soil erosion is socially constructed, for example, appears to deny the physical forces and processes that determine soil movement, which are usually the purview of soil scientists and not critical theorists.

Debate about constructivism in science revolves around several specific and somewhat irresolvable philosophical disagreements. It is impossible to fully review these here. Nevertheless, these are important for understanding why many political ecologists make constructivist claims. These boil down to a basic suspicion on the part of many political ecologists that the categories of reality described in much environmental science and state management are ultimately arbitrary and serve specific, often narrow, political interests. Constructivists argue that categories (indigenous, scientific, or otherwise) may adequately capture some commonalties in the pattern of reality but they are no more accurate than any other possible classification. Any given classification clusters and excludes different phenomena, but does so in a no more accurate way than its alternative. "Scientific" expertise only lends more social and political weight or credibility to one arbitrary classification over another. As Foucault asks:

> when we establish a considered classification, when we say that a cat and a dog resemble each other less than two greyhounds do, even if both are tame or embalmed, even if both are frenzied, even if both have just broken the water pitcher, what is the ground on which we are able to establish the validity of this classification with complete certainty? On what "table," according to what grid of identities, similitudes, analogies, have we become accustomed to sort out so many different and similar things? (Foucault 1971, p. xix)

Consider the classification of a species of palm tree or fish. Are palms trees? While they are in some cultural and scientific lexicons, they are not in others (Ellen 1998). The identity of catfish and the inclusion and exclusion of Asian varieties from its classificatory domain have been a matter of US congressional testimony (Mansfield 2003).

Constructivists seek to highlight this contingency in their assessment of environmental science and planning. In so doing they suggest that current facts (or those asserted at any historical point) are not inevitable outcomes of empirical inquiry, that the natural world can be described in a range of categorical fashions beyond those that currently exist, and that this decade's scientific "truth" is apt to change with the political and social wind. Many hardcore "realists" adopt the opposite position. Most practitioners in political ecology dwell somewhere in between.

In application, this approach reveals much. Consider the case of West African deforestation. The universal account of forest conditions throughout all of Africa, and especially in West Africa, has long been that tree cover has been declining at an accelerating rate in the last few years. Official hard facts and statistics to build this case have been promulgated by the Food and Agriculture Organization (FAO), the World Resources Institute, and a range of other credible expert sources. The blame for all this tree cover loss, in most accounts, lies with poor, ignorant, and overpopulated local communities – the traditional targets of apolitical ecology (Fairhead and Leach 1998).

As James Fairhead and Melissa Leach suggest in their extensive exploration of data from colonial policy, contemporary development narratives, official statistics,

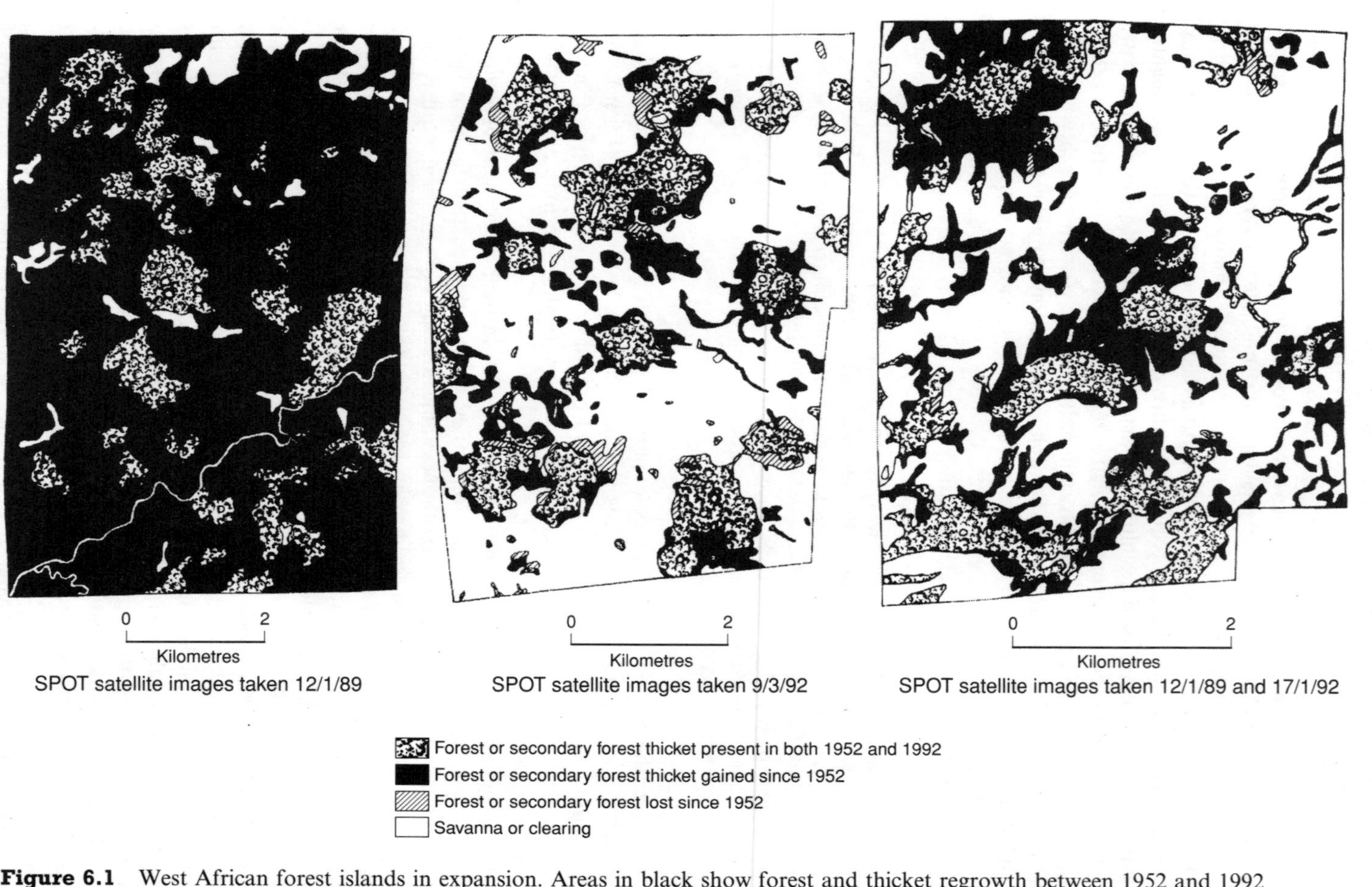

Figure 6.1 West African forest islands in expansion. Areas in black show forest and thicket regrowth between 1952 and 1992 (Reproduced from Fairhead and Leach 1996, figure 2.2)

oral histories, and air photography, the case for regional deforestation is indeed remarkably thin. When forestry officials look at the islands of forests around villages and imagine that they see remnants of what used to be larger forests, in fact they are seeing forests actually in expansion (Figure 6.1). This reforestation and afforestation occurs, moreover, specifically as a result of informal local land management by village producers, not despite them (Fairhead and Leach 1996).

What does this suggest about the commonsense narratives that supported the opposite claims? From a simple realist perspective, they were "wrong"; bad science produces bad numbers. Political ecologists, however, in keeping with their constructivist urge, want to ask further questions. How did this conception of environmental change persist, and indeed survive into the present decade, in the face of its readily recognizable falsehood? What social and political benefit might it have had for conservation experts, government officials, and international agencies? Why and how did it get constructed the way it did? To whose benefit and whose loss? And even where some degradation is occurring, is it an irreversible disaster or rather a temporary change of ecological state? If the latter, what drives a narrative of crisis?

Thus, for many important questions in political ecology, the issue of environmental destruction is not the only relevant factor in determining what happens in the world. Control over resources is commonly not adjudicated by whether overgrazing, erosion, or biodiversity decline is actually occurring in the landscape, but rather by the *accounts* of environmental conditions and change that are held as true by decision-makers, local people, and competing interests. Moreover, since the scientific practice of determining land degradation occurs within a politicized environment, formal land degradation study does not provide by itself an absolute and neutral position from which to adjudicate disputes over environmental control and management.

Hard and soft constructivism

But to say that environmental facts like soil erosion in East Africa are constructed, rather than inevitable, inherent, or stable, is still to underspecify how political ecologists think. In fact, there is a range of possible "commitments" to the constructivist position in political ecology and by no means is there simple agreement among researchers as to which should prevail. The multiple forms of constructivism, drawing on their elucidation by Demeritt (1998) and Sismondo (1993), include what I call here "hard" or "radical" constructivism, and "soft" social object and social institutional constructivism. Each makes a different claim about how science interacts with other social practices and each provides a different mandate for how to treat expert claims.

"Radical" constructivism The environment is arguably an invention of our imagination. What we know from experience of much of the world, moreover, is related to us through stories, conventions, and idea systems that we learn from other people. Processes and transitions are captured in conceptual terms that are fundamentally symbolic and abstract. This is as true for modern urban residents as it is for forest-dwelling shifting agriculturists, perhaps more so. Ideas about nature inevitably reflect our social world.

In its most radical form, "hard" constructivist epistemology takes this symbolic and ideational character of environmental knowledge extremely seriously, insisting that it is social context alone that conditions and determines our concepts for understanding the world, and so *creates the world*, at least effectively, in the process. This position suggests that things are true because they are held to be true by the socially powerful and influential, because they are true on television, and because they are true in our minds. This radical position is relativistic insofar as it holds that science, as one specific social method, cannot be used for adjudicating disputes between different claims about what is real, all of which are arbitrary. As philosopher of science Steve Woolgar insists, "nature and reality are the by-products rather than the predeterminants of scientific activity" (Woolgar 1988, p. 89). Environmental conflicts are, therefore, struggles over ideas about nature, in which one group prevail not because they hold a better or more accurate account of a process – soil erosion, global warming, ozone depletion – but because they access and mobilize social power to create consensus on the truth.

For most political ecologists, this approach is somewhat too sharp a double-edged sword. While it allows a critical examination of how politically empowered environmental science has influenced and created the environments of the world around us, which is an important political ecological project, this approach disallows the reference to non-human actors and processes (like soil, trees, and climate) in explaining outcomes, making it unattractive to many researchers. While producing a valuable open space for accepting and appreciating alternative constructions of the environment held by other social communities, like forest dwellers, nomadic herders, and religious philosophers, this approach makes the symbolic systems of humans sovereign over all other reality, apparently disabling empirical investigation in traditional environmental science.

"Soft" constructivism As a result, most political ecologists tacitly cling to a "softer" form of constructivism, which holds that our concepts of reality are real and have force in the world, but that they reflect incomplete, incorrect, biased and false understandings of an empirical reality. In other words, the objective world is real and independent of our categorization but filtered through subjective conceptual systems and scientific methods that are socially conditioned. Within this approach to constructivism, there are differing emphases, which center attention either on people's misunderstanding of objective facts or on the social biases that enter into scientific exploration (Demeritt 1998).

In the first case, false and socially biased categories of the world, like "race," are important to understand and explore even while their reality – consistent racially differentiated genetic differences – does not objectively exist (Mitchell 2000). Since people hold them experientially, these concepts, or social constructions, make a difference in the world, often with pernicious effects, and therefore need to be understood. This "social object" approach to nature is attractive for political ecologists, who are able to assume that ecological science can reveal real environmental trends, like soil erosion, while social investigation can show how ignorant people can create false pictures of the world, like "desertification," through power-laden social processes. This approach is satisfactory for most researchers since they consider themselves scientists (or at least allied with scientists). They can insist that their way of seeing the problem, using the tools of science, helps to unmask biased and incorrect views of nature.

The confidence that such an approach places in scientific practice, however, is highly problematic. As radical constructivists persuasively point out, and as is revealed in histories of science, the very categories of scientific investigation are the same order of "social object" as the false commonsensical notions of the lay population.

The history of ecology is revealing in this respect. The dominant theories of the operation of natural systems have consistently reflected the prevailing social languages and assumptions of their times. Emerging during the high industrial age, the science of ecology came to depend heavily on metaphors and concepts from mechanical engineering, with orderly, cyclical, processes structured around balance and symmetry. It also drew heavily, and somewhat contradictorily, upon philosophical Romanticism and the obsession with holism and interdependence, as is found in Romantic writers like Henry David Thoreau (Worster 1985a). These metaphors, on which science depends, became unsatisfactory in recent years, either because they reflected reality poorly or didn't fit changing social and cultural codes, and now are in a state of upheaval.

This should be in no way surprising, ecologist Daniel Botkin insists; previous views of nature, either as an organic whole or as a divinely ordered house, clearly reflected the social languages available to those who sought to explain nature's order (Botkin 1990). So too, the history of primatology, studied in careful detail by Donna Haraway, shows similar socially bounded evolution; the changing topics of explorations and experiments on chimpanzees and gorillas (maternal instinct, aggression, competition) reflect the social concerns of their historical moment, reading more like a history of contemporary American culture than orderly evolution of animal ethology (Haraway 1989). Our scientific ideas of nature inevitably reflect the social conditions and dominant metaphors in which they were formed. Nor is this necessarily bad. With changing metaphors come emerging ways of thinking about and reinventing the world. Even so, science is not free of "social objects."

An alternative soft constructivist approach, "social institutional constructivism," allows that such biases are a structural part of scientific practice, but that they nevertheless do not solely determine the conditions of the objective material world. Rather, these conceptual biases in science help to explain why science sometimes gets facts wrong. For social institutional constructivists, wrong ideas about nature are a product of the inevitable "socialness" of scientific communities. Over time, however, and through progressive experimentation and refutation, the "social" ideas are purged from our understanding of nature, moving towards a true understanding of the objects of the natural world. This is especially true, a social institutional constructivist might argue, as contemporary ecology and life sciences become more and more reflexive about the metaphors that underpin their analysis of objective systems (Sullivan 2000).

As an approach to political ecology, this is perhaps the most common and attractive epistemological compromise. Knowledges are all different, most researchers maintain, and different experiences, like those of biologists, herders, historians, farmers, and foresters, do indeed produce extremely different categorical structures for interpreting the objective realties of the natural world. Even so, these knowledges can be adjudicated by incorporating local ways of knowing into a flexible but rigorous scientific framework, which will distill myths from realties and produce better, more emancipatory knowledge (see especially Batterbury et al. 1997; Sullivan 2000). Acknowledging the socially situated character of science, the method can still be used to test contested claims (Forsyth 1996).

This approach is a pragmatic compromise but is troubling for many observers of science and politics. From a philosophical and historical point of view, it is somewhat unconvincing and asymmetrical; social institutional constructivism insists that only falsehoods, those situations where scientific facts are wrong, can be explained socially, whereas facts and true understandings of nature have no social component. Following science studies researcher Bruno Latour, under such an account: "Error, beliefs, could be explained socially, but truth remained self-explanatory. It was certainly possible to explain belief in flying saucers, but not the knowledge of black holes; we could analyze the illusions of parapsychologists, but not the knowledge of psychologists; we could analyze Spencer's errors but not Darwin's certainties" (Latour 1993, p. 92).

For some political ecologists who are most definitely interested in how environmental concepts become powerful and true, this might be quite unsatisfactory. Such an approach only functions to explain things that we know to be "wrong," including the dominant account of nature, and only if we are already confident that whatever the facts are, they are wrong, and scientifically untrue. Generally this means that the claims of others ("enemies" like state soil conservationists, World Bank officers, or seed company representatives) can be disposed of as *constructions*, while the claims of other parties ("allies" like local herders or fishermen) are vindicated as holding environmental *knowledge*. Where even those knowledges fail the practical tests of science – whatever that is taken to mean – they too become constructions.

The ethical implications of such an approach are therefore equally problematic. Many political ecologists (though by no means all) come from Anglo-American universities and think tanks, travel on relatively large budgets, and exercise tremendous institutional authority. To arrive in other contexts, whether woodlands in Alabama or pastures in Mongolia, and consider it appropriate to provide adjudication between competing local claims should quite readily be construed as the height of colonial arrogance, rightfully denounced by postcolonial and subaltern critiques of academic research enterprises (see Chapter 3). Such an approach does little to dethrone the very structures of hegemonic power that political ecology seeks to challenge.

Constructivist claims in political ecology

Whatever type of constructivism is used, the central claims in the field follow several common threads. They seek to show how ideas and narratives about nature and society are mobilized in environmental struggle. The following represents a sample of such arguments with a few illustrative examples.

One common argument is that many things that are by no means environmentally natural are made to appear that way, and vice versa. In perhaps the most well-known and controversial case, environmental historian William Cronon, after examining the changing meaning of the concept of "wilderness" in Western history, concluded that the idea is historically contingent. Given the implication of humanity in producing "natural" environments all around them and the presence of natural processes in non-wilderness areas like the city, wilderness must be viewed as a social construction, and one that actually bars effective management and conservation, placing humans outside of nature as it does (Cronon 1995). The idea of wilderness, therefore, and the invocation of the pristine in wild nature is, by implication, less a reference to a real condition than it is an emotive image with broad political effects, including the promulgation of conservation reserves across the world, where traditional local residents are excluded.

Cronon's conclusions were far from uncontroversial, however, and many observers argued back vociferously that wilderness was "real," accusing Cronon of undermining progress in environmentalism. Opponents stressed the degree to which Cronon's own discourse – there is no wilderness – might be used by anti-environmental and economic development advocates to promote reckless exploitation.

Nevertheless, the political and ecological implications of this line of thinking have proven useful in progressive research around the world. Neumann (1998), for example, has carefully documented the way in which imported Anglo-American wilderness aesthetics – ideas of how wilderness *ought* to look – were imposed on African landscapes, inventing environments that had previously not existed. Labeling this aesthetic natural politically facilitated the removal and disempowerment of local people who had participated in creating the very "natural" landscapes of the tropical and subtropical savanna that colonial and postcolonial officials sought to preserve (see Chapter 8).

Similarly, though in an inverse fashion, political ecologists have sought to demonstrate the way in which environmental "problems" are constructed where none exist, or at least where the "problem" is a product of largely ecogenic processes. Crisis representations in particular, where environmental situations are framed as unprecedented and disastrous, have been politically useful for international agencies seeking funding (Jeanrenaud 2002), even while the long-term trajectory of such system changes may be highly variable and unstable (Behnke and Scoones 1993).

Reverse arguments are also possible, of course, where socio-politically caused outcomes are described as "natural" and therefore apolitical. Indeed, the fundamental anti-Malthusian critique in most political ecology is an attempt to expose the way some observers *construct* economic and demographic tragedies (e.g., mass starvation and disease) as natural by-products of overpopulation (Harvey 1996).

Political ecologists thus commonly examine the way claims about environmental systems become rooted in the political-economic systems that produce and sustain them. Such arguments diverge somewhat in tone. Some political ecology of forestry, for example, explores the deliberate and systematic way forest conditions are recorded by state agencies (Kummer 1992; Bryant 1996). The resulting official records of land cover change or degradation are commonly overstated or understated to divert attention away from a problem or, alternatively, to capture resources for solving a problem that may not exist. This approach, which emphasizes the conscious manipulation of environmental statistics and representation, might be called a *rhetorical* or *tactical* approach to construction.

A more definitively constructivist approach usually emphasizes the non-conscious way in which state managers, local people, and international agencies hold different normative ideas of the environment. Such an approach puts less emphasis on the intentional and strategic use of ideas and narratives about nature, and is more focused on how "naturalization" occurs, highlighting the social process whereby the constructedness of environmental concepts and practices is forgotten (Robbins 1998b).

"Barstool" Biologists and "Hysterical Housewives": The Peculiar Case of Local Environmental Knowledge

If expert accounts of nature are implicated in political struggles and represent constructed ways of viewing nature, it is logical to ask whether local, non-expert accounts are more accurate and practical. The resulting branch of constructivist investigations in political ecology explores "local" or "situated" knowledges.

Such investigations are indeed very old. The cultural ecology of environmental knowledge, starting in Kropotkin's time and advancing into the environmental anthropology and geography of the 1960s (Chapter 3), established a record on the accuracy and practicality of local ecological practices, especially amongst traditional people practicing subsistence production. This tradition continues into the present with empirical work revealing the strength and efficacy of traditional ecological knowledge on multiple levels, including (following Berkes 1999):

- immediate empirical knowledge and taxonomies of plants, animals, and soils
- practical knowledge of functional relationships and processes, like ecological succession
- social knowledge of traditional rules, institutions, and systems of management
- conceptual systems, worldviews, and more abstract beliefs that order experience and interpretation of the environment

Whether or not traditional forms of environmental knowledge are exceptional, different, or superior to laboratory knowledge or that of experts more generally, however, is a matter of more general debate. Certainly the failures of many imposed environmental management solutions around the world, ranging from failed crop introductions to disastrous property regime changes, highlight the appropriateness of environmental knowledge developed locally (Brokensha et al. 1980).

Customary tenure relationships in Southeast Asia, for example, which recognize distinct land types for sedentary cropping, shifting cultivation, hunting, and gathering, have historically functioned extremely well until replaced by certain forms of imposed freeholding and land marketing with little acknowledgment or linkage to local environmental systems (Cleary and Eaton 1996). New England fishers similarly show a sensitive and well-developed spatial conception of fish biology – including explicitly mapped knowledges of breeding grounds, fish migrations, and other dynamic characteristics of the undersea environment – far superior to the aspatial conceptions of scientific bioeconomics most commonly used in official management (see Chapter 8) (St Martin 2001). Indeed, where local knowledge of biodiversity has been shown to be well developed, the most serious concern is not its efficacy, but rather whether local people will receive control over and due compensation for that

Box 6.1 Fikret Berkes's *Sacred Ecology* between two worlds

Traditional systems of environmental knowledge are often awe inspiring in their sophistication. Such knowledge can be something simple, like Cree Indian fishing practices, which set and alter fish net size to respond to changing harvest rates amongst differing age structures of fish populations. Or it may represent something considerably more complex, like the organization of the complex *ahupua'a* farm and fish water management systems of the Hawaiian islands, which historically ran from mountain slopes to the ocean, integrating water flow through farmlands, down through brackish fish ponds for harvesting marine foods, outwards through forest belts used to protect land from storm surge, and on towards the sea. The incredible effectiveness and wisdom of such traditional knowledge is a long-acknowledged fact of academic cultural ecology.

Fikret Berkes's *Sacred Ecology: Traditional Ecological Knowledge and Resource Management* represents more than simply an attempt to codify that local knowledge or to create a systematic account of how it works. Instead, Berkes's comparatively terse volume represents an effort to reform formal, reductive, and model-driven ecological science so that it embraces the sacred elements of traditional science. It is a manifesto aiming to reconcile reduction with integration, technology with wisdom, and skeptical inquiry with a feeling for the sacred.

The project was borne of Berkes's useful if sometimes awkward position between two worlds. Trained as a marine scientist and applied ecologist in the early 1970s, he turned down an opportunity early in his career to work in what he recently described as "reductive" formal marine ecology, choosing instead to work with an anthropologist colleague and spend his time fishing with the Cree Indians of James Bay, something he would continue to do for the rest of his long career. As a result, Berkes found himself at home collecting quantitative fish population data while simultaneously collecting folk histories of the sacredness of animals. The book reflects this vision, seeking to show not only that indigenous ecological knowledge "works" in the objective sense, but that it is spiritually whole, a unity of mind and nature.

The politics of this reconciliation between scientific and sacred ecology, however, are less well defined or discussed. The book provides a wealth of examples where modern scientific knowledge systems become tools for the erasure not only of local people's knowledge but also of their control over resources. But Berkes is less interested in discussing these political processes than in simply documenting and defending the knowledge of historically marginalized people. This reflects his own training and influences, far closer to marine science and cultural ecology than to the agrarian economy and Gramscian peasant studies of most political ecologists.

This decision makes the book more seed than hatchet, showing how local knowledge could help to heal modern ecology, without fully addressing and criticizing the political and epistemological barriers that make that outcome unlikely. Even so, Berkes remains one of the most experienced and articulate translators of traditional environmental wisdom, and brings a depth of experience sometimes lacking in political ecological explanation.

knowledge when it is appropriated by plant growers and pharmaceutical companies (Brush and Stabinsky 1996).

Even so, sustained consideration suggests that highlighting distinctions between local and scientific knowledge obscures more than it reveals (Agrawal 1995). Other research has highlighted the differential value of local and scientific knowledge in varying contexts, as in rural Mexico where local knowledge has been shown to be somewhat less effective at evaluating the medium-term impact of human actions than scientific knowledge, even while being far more flexible and adaptive in its implementation (Klooster 2002). Similarly, research has highlighted the adaptability and persistence of local knowledge as it articulates with modern management systems, even in the face of globalizing pressures (Brodt 2001).

Attempts at integration of environmental management regimes and local knowledge systems are also therefore increasingly apparent, as in Senegal where local histories of ecological succession are incorporated into fire and plantation planning (Lykke 2000), or in Lebanon where local knowledge of mountainous terrain has been incorporated into land use mapping to develop otherwise unavailable data and facilitate democratic participation in planning (Zurayk et al. 2001). The construction of nature by officials and locals is sometimes well integrated and there has been a recent call for "hybrid research," which evaluates the usefulness of local knowledge based on a yardstick of practical efficacy (Batterbury et al. 1997).

More commonly, official and scientific managers continue to dismiss local environmental knowledge as politically interested, not objective, and poorly informed, even and especially in the first world. In the environs of northern Yellowstone, for example, state ecologists commonly dismiss the mental ecological models of hunters as "barstool biology." In Fernald, Ohio, the concerns of local women observing adverse environment hazards in local water and air are characterized by scientists and planners as those of "hysterical housewives" (Seager 1996). As Fikret Berkes explains, these accounts, informed by local experience and opposed to the imposition of control over local resources, represent a "challenge to the dominant positivist-reductionist paradigm of Western science," largely for reasons that "have to do with power relations between Western experts and aboriginal experts" (Berkes 1999, p. 11).

So while local knowledge is increasingly on the agenda, the difference between formal and informal knowledge systems remains a source of conflict. And while constructivist accounts in political ecology can and must acknowledge the interested and contextual character of local knowledge, they must also explain the structured biases built into official knowledge systems, which are used by experts to secure employment, control resources, and justify extraction and enclosure (Robbins 2000). The knowledges of scientific practitioners and other "experts" are embedded in cultural norms, social relationships, and value-laden judgments, even and especially in large-scale scientific investigations like climate change research (Demeritt 2001).

The case of local environmental knowledge is, therefore, an important and pressing one for political ecologists, who must explain how certain accounts of environmental process became dominant and to what effect. Why and when do expert accounts of land degradation come to crowd out local accounts, such that some local environmental practices, like the use of fire in land management (Kull 2000) or the practice of swidden agriculture (Dove 1983), are singled out for restriction and control?

Eliciting environmental construction

To study the construction of nature is, however, as difficult as studying land degradation and amelioration, and presents equally complex methodological problems. Since knowledges and constructions span the scale from local taxonomies and narratives, to conceptual symbols and metaphors, empirically studying them is a challenge.

Talk and text: construction in discourse Constructions of the environment are communicated in myriad media, including advertisements, folk songs, photographs, scientific documents, daily conversations, diaries, and landscape paintings. Constructions are rarely fully embodied or realized in a single form, moreover, and are joined together from a collection of parts. Indeed, a construction of the environment (or more generally a "discourse") represents a combination of "narratives, concepts, ideologies, and signifying practices" (Barnes and Duncan 1992, p. 8), including the things people both say and do.

In this sense, constructions of the environment succumb to many modes of analysis, but by the same token require many methods to be revealed. Political ecologists commonly scour old forestry records, conduct open-ended interviews with producers and managers, read lengthy government reports, and even examine commercial advertising for clues about how, when, and why the environment is constructed through social and political processes.

In exploring divergent constructions, moreover, like those between herders and farmers, men and women, or the rich and poor, research typically seeks a broad and representative range of sources. Unfortunately, not all communities and individuals, especially those in differing positions of social and economic power, communicate their interpretations of nature in the same way. Such differences in modes of communication mean that the constructions bound up in those differing discourses are sometimes not treated on equal terms. While the discourses of scientific ecologists studying rangeland dynamics, for example, may be embedded in statistical documents, those of local pastoralists may rest in oral histories, herding practices, and place names. These are hard to place on a common scale. Comparison, discussion, and analysis is therefore difficult.

Categories and taxonomies In many cases, constructions of the environment are most explicitly reflected in the categories with which nature is described and ordered. Local people, scientists, and other observers all have taxonomies for soils, species, and land covers, which both drive and reflect their constructions of the environment.

Eliciting such categories is a matter of intensive survey, where research seeks to produce exhaustive lists of the conceptual differences between groups of trees, landscapes, or soils, drawing upon many traditional techniques in environmental perception research as well as cultural ecology and ethnobotany. It is also fraught with linguistic pitfalls and problems of interpretation. Many taxonomies of plants or animals, like the Western/scientific Linnaean system, are organized in complex hierarchic fashion, meaning that careful listening is required to determine the lumping and splitting of environmental phenomena, and many species and landscapes have multiple names (Berkes 1999). Most problematically, the classification of ordinary things is often so taken for granted, by farmers, scientists, or consumers, that it is easy to misunderstand the purpose and direction of conversations directed towards weeds, laboratory equipment, or other daily objects.

The benefits of careful study, however, are many. The list below shows the categories of land types amongst the Ifugao of northern Luzon in the Philippines drawn from years of fieldwork by H. C. Conklin. Each of these categories reflects a mix of land use and land cover and shows the different conceptual divisions between various components of human-managed production landscapes. Most prominently, the category system is centered around processes of succession and ecological state transitions in which "natural" landscapes like forest and grassland over time cycle through a system of swidden (slash and burn) production and secondary growth. The system not only reveals the conceptual world of the Ifugao, it further shows the way they make a living, the way they manage and respond to environmental change, and the divisions in property and production in their communities.

Ifugao categories of land types and succession patterns (following Ellen 1982 and Conklin 1968)

Category	*Description*	*Species*	*Succession to*
Mapulun	Short, low, open grassland	*Imperata* spp.	Mabilau
Qinalahan	Public forest	*Thermeda* spp. (mid-mountain climax)	Habal
Mabilau	Cane grassland and secondary growth	*Miscanthus* spp. (canegrass)	Qinalahan
Pinugu	Private forest grove	Timber and fruit trees, erect palms, rattan	Payo
Habal	Slope swidden fields	Sweet potatoes, taro, yams, manioc, corn, millet	Pinugu
Lattan	Residential hamlet terrace		
Qilid	Drained terrace	Sweet potatoes, legumes	
Payo	Pond field, rice terrace	Rice and taro	

No data is available for the categorical systems that land management officials or experts from the Philippine government apply to the same landscapes, but it is certain that they would reflect little of the subtle succession and production variations evident in the local taxonomy. What might this mean for control of land and the trajectory of its development? What if, for example, portions of Mapulun grasslands and Qinalahan forests were enclosed under conservation mandates or wilderness preservation efforts, specifically because in official taxonomies these became "pasture" and "forest" – atemporal, permanent, and fixed land cover types? In that case, key succession spaces to secondary growth and swidden fields would be lost, creating bottlenecks in production, threatening livelihoods, and creating scarcity in resources. Categorical and taxonomic information provides the building blocks for a political ecology of landscape production and control.

Spatial knowledge and construction The categorical lumping and splitting of the natural world is also an inherently spatial process; the maps of production, degradation, and control that we carry around in our heads reflect deeply rooted and socially influenced constructions of nature. These also provide methodological opportunities to explore political ecology.

Most of the methods for deriving these spatial constructions draw heavily on the research tradition of "cognitive mapping," which encourages people to map their surroundings (Kitchin and Frendschuh 2000). Growing out of these practices, more recent work has tried to codify local conceptual geographies as Geographic Information System (GIS) data. These efforts at "geomatics" can later be used to defend native land rights against the depredation of lately arriving settlers or state authorities bearing formal titles (Poole 1995). Counter mapping, the most politically explicit form of geomatics, has as its aim to "appropriate the state's *techniques* and *manner of representation* to bolster the legitimacy of 'customary' claims to resources" (Peluso 1995, p. 384, emphasis in original).

The challenges in this methodological area are many, however. Paper and pencil mapping is foreign to many extremely knowledgeable local or traditional people, who may be far more comfortable in the oral communication of geography, or with sticks and rocks laid out on the ground. So too, the "formal" cartographies arranged by professional mapmakers following standardized guidelines may not reflect the environmental geographies in the heads of the individuals or institutions that produced these maps. The asymmetry between different communities – those historically called experts versus those historically called lay – may in fact be reinforced through poorly defined cognitive exercises and sloppily executed mental mapping.

Narratives of ecological process and change Environmental constructions are not limited to detailed taxonomies and geographies of production. Some of the most obvious and important constructions, especially those that impinge on the political control of the environment, are embedded in the stories of environmental change and memories of past ecologies that people hold. In explaining how environmental conditions change and why, people not only articulate their notions of how ecosystems work and the patterns of cause and effect they perceive, they further reveal their perceptions of how landscapes looked and functioned in the past.

Such narratives are usually rooted in collective agreement and tacit consensus reached within communities – whether these are peasant villages, planning offices, or GIS labs – such that stories of change provide a window onto collective priorities and group memory. When, for example, West African colonial and postcolonial land use planners reach an oral and written consensus that deforestation is serious and ongoing, despite some evidence to the contrary, we are forced to consider the structures and systems of agreement that allow that idea to prevail (Fairhead and Leach 1996).

At the same time, lack of consensus is also revealing in political ecology. In Bolivia, for example, Zimmerer (1993) derives dramatically divergent accounts of the status and causes of soil erosion in agrarian environments. Narratives that hold peasant ignorance, changing agrarian practices, and transnational exploitation accountable for erosion are held by development institutions, local producers, and trade unions, respectively. Such analysis underlines the way rifts in environmental interpretation follow existing political divisions.

The pitfalls of narrative approaches are several. Individuals and communities do not always believe what they say, and belief does not always lead directly to predictable action. Depending solely on the environmental narratives of individuals or groups is not a clear and open window into the complex constructions of nature held within and between groups. Rigorous techniques, survey methods, and sustained presence in a community are all prerequisite to clarifying and verifying human models of environmental change. Even so, such work is essential for robust explanations of the causes and consequences of ecological and political transition.

Genealogies of representation: environmental history While all of these approaches are revealing and powerful, the most sophisticated readings of environmental knowledges, narratives, and imaginaries require deep historical analysis. As noted previously, such historical analyses are ultimately necessary to shed light on the moments of invention or transformation that fix what appear to be timeless concepts to historical moments of political and economic change.

In perhaps the most trenchant recent analysis of this sort, Willems-Braun (1997) surveys contemporary representations of forest and wilderness in British Columbia, puzzling over the ways in which the environment, though one contested by contemporary native peoples, is represented in public debate as a "purified space" devoid of native presence. While such a representation is of course politically convenient for the contemporary Anglo-Asian residents of the region debating the future of the forest, such an instrumental answer is ultimately unsatisfying for Willems-Braun, since it explains neither the power nor the longevity of this construction. Excavating early documents on the region by explorer George Dawson, his analysis uncovers the depth of the "purification" of the ecological record and the way in which even early writing expunged the native cultural presence from the wilderness of the forest. This lays a powerful groundwork for a contemporary debate that, though it acknowledges native peoples, allows them little voice in the control of "wild spaces," which is ceded instead to expert ecologists and foresters (Willems-Braun 1997).

In another example, Sullivan (2000) examines the roots of the contemporary view of uncontrolled land degradation in Namibia, which is commonly offered as justification for restricting use of the land by local residents. The research burrows into the past, including accounts by eighteenth-century travelers, reports by late-nineteenth-century managers, and studies by ethnographers in the mid-twentieth century. These diverse texts all consistently report conditions using the same terminology and the same mental pictures, which together invoke a scene of overgrazed and eroded soils. The depth and persistence of these stories is, at least in part, both a cause and consequence of their power in the planning process, and contributes to limiting the power and access rights of local people.

These genealogies are increasingly a part of political ecology. The work is far from simple, however, and the key linkages which connect the deep histories of explorers or colonial officers to contemporary politics today are more often inferred than demonstrated. This is largely because records are sketchy, partial, and littered with contradictory and opaque evidence.

Despite these methodological challenges, a serious and rigorous engagement with the construction of the environment benefits from tracing contemporary claims about nature backwards to their roots.

Box 6.2 Vanishing natives and other postcolonial magic tricks in Braun's "Buried Epistemologies"

Postcolonialism is a big word. It also sometimes seems to belong to an obscure and arcane world of "critiques" and "theorizations" with little use for explaining things like trees. As Bruce Braun demonstrates in "Buried Epistemologies: The Politics of Nature in (Post)colonial British Colombia" (Willems-Braun 1997), however, the term captures much to explain what is going on in the forest.

In his investigation of conflicts over control of "wilderness" areas around Cloyoquot Sound, a long-occupied but heavily forested place in western Canada, Braun turns to the language of environmental groups and scientists, as well as to nature writing and photography, to show the way foresters, environmentalists, and other powerful groups think about and represent both nature and native culture. He argues that these groups define the terms of struggle – "nature" versus "culture" – in such a way that the indigenous peoples in the region, the Nuu-chah-nulth, are written out of the history of the production of the Cloyoquot landscape, except insofar as they are "traditional" people, living amidst totem poles and paddling canoes. The result, of course, is that the claims of these people over the forest and its fate are largely eclipsed.

More than this, Braun demonstrates that this "habit of thinking" (epistemology) is something that the contemporary Anglo majority – both industrial foresters and environmentalists – inherited from colonial logic and practice, with its methods of classifying, recording, and describing the world. He draws upon photographs and journals from a surveyor, George Dawson, who in the 1870s recorded the Cloyoquot landscape in such a way that native peoples fit neither as part of the natural environment (being cultural rather than natural) nor as part of the emerging Canadian national polity (being traditional rather than modern). The somewhat depressing conclusion is that violent exclusionary systems of domination are persistent and commonly reproduced even by "liberal" environmental observers with a purported sympathy for native peoples. They are made to "vanish" by a trick of epistemological habit rooted in the privilege of the colonizer, which ultimately determines what kinds of land uses can occur in the forest (preservation and timbering) and delimits who gets to say so (Anglo environmentalists and foresters).

For political ecologists, whose stock-in-trade is mostly writing and telling stories about people like the Nuu-chah-nulth, this serves as a warning shot across the bow. With its roots in cultural ecology, political ecology also has many habits of thought; those buried epistemologies that tend to make certain categories (e.g., peasants, nomads, old growth forest, etc.) "real." These categories are inherited, however, often from very ugly systems of exclusion and domination.

More than this, at the time of its publication, "Buried Epistemologies" suggested for many researchers, this author especially, that systems of representation, like photographs, journals, and the scientific categories, are as fruitful a place to understand environmental politics as in fields, factories, and workshops. So for political ecology Braun helps to open the door to a crucial renaissance of culture in the field.

Methodological Issues in Political Analysis of Environmental Construction

In sum, the frameworks with which we imagine the non-human world are as important (and contested and puzzling) as the variability of that world as understood in ecological science. The mandate in much political ecology, therefore, is to map the politics of environmental ideas as carefully as the politics of material ecological change, working to link the two across space and time.

There are several immediate methodological barriers to effective analysis. The way in which knowledge systems are communicated and recorded can lead to asymmetrical analysis. The complexity between spoken and unspoken reasoning opens the door to confused attribution of motives and politics. These problems are often coupled with poor sampling strategies on the part of many political ecologists studying knowledge. Who, precisely, is interviewed or surveyed? Do they represent common knowledge? Are they experts? How does knowledge vary within populations? A general lack of attention to specific methodologies in local knowledge studies remains a serious problem (Davis and Wagner 2003).

Despite these difficulties, however, construction is a crucial process that defines, channels, and makes manifest struggles over the environment. As an empirical project, researchers must:

- elicit the conceptual vocabularies of the range of participants in ecological process and struggle
- determine the relationships of rhetorical and deeply discursive formations to environmental and political practices
- seek methods that assure the symmetry of inquiry between official knowledges, often in elite languages and formal texts, and local ones, which are often transmitted orally and in local vernacular
- explore the way environmental narratives and cartographies unite and divide communities that might otherwise seem disparate or unified
- establish the roots of the most obvious and taken-for-granted environmental conceptions that drive, direct, and dominate conflict

As is the case for ecological analysis, political ecologists have not always been entirely attentive to these methodological imperatives. Again, however, this reflects the complexity of discursive/material interactions more than the research failings of the field. As we shall see, the sustained interrogation of these issues, construction and destruction, environmental imaginaries, and ecological impacts, do serve as the center of contemporary research into the politics of the environment.

Part III

Political Ecology Now

In which four overlapping arguments in contemporary political ecology are surveyed, their relative merits weighed, and some nagging problems discussed. Herein we also discover that doing political ecology requires patience, imagination, and a willingness to sometimes fail.

7

Degradation and Marginalization

Ghana entered a program of structural adjustment in 1983, planning to bring the unfettered power of free trade to this largely agrarian country by opening its borders and capitalizing on the export of cash crops to a global market. Under reform programs that reduced tariffs, eliminated price controls, and set foreign exchange on a global market, productivity was expected to increase in agriculture by some 4 percent per year as farmers made capital improvements, paid for with receipts from sales of oil palm, cocoa, cashew, and other crops. The optimism of the World Bank and the Ghanaian government was based on strong precedent. Ghana had entered the independence era with a wealth of natural resources, an educated population, and a well-developed infrastructure.

But in the 15 years that followed, the overall rate of growth in agriculture was a disappointing 2.5 percent. Indeed, while some farmers were able to seize control of investment opportunities, most were not, and many enterprises continued to struggle, especially under a surprising labor shortage. After enduring the hardships typically associated with structural adjustment, including decreased employment and increasing poverty, migration, and hunger, Ghana had little to show for having swallowed the medicine offered by the world community.

Apolitical explanations of this failure, and failures like it, take many forms. Market reforms, one might argue, take time to work and need to be fully realized; slow adjustments are inevitable – 15 years is simply too short a time span to gauge success. High population growth in Ghana (around 3 percent per annum) might also be argued to absorb surpluses and stunt productivity. As we have seen already, however, such explanations tend to be anti-contextual, superficial, and non-processual. They also do little to diagnose the specific political and ecological drivers that regulate change, and thus have little practical use.

Examining the problem in terms of regional and local struggles and ecological transformations might result in a different explanation, however. Approaching the problem in this way, from the ground up, Louis Awanyo offers an alternative account. Interviewing several hundred farmers in the Berekum region of Ghana, Awanyo searches for the processes and system elements that restrain agrarian growth. After long-term ethnographic work in these rural communities, he suggests that the answer lies in the relationship between an aggressive weedy plant and the labor structure of a stratified rural society.

Specifically, his analysis shows that the biggest problem in local production is the control of a thick-growing, fast-germinating weedy plant, *kra wo ni* (*Chromolaena odorata*). This highly aggressive invasive species, probably native to the Americas, spreads at a rapid pace, overtaking farm fields in a single season. It therefore represents a massive regional barrier to production, and farmers and planners in rural Ghana understandably view its spread as a form of land degradation.

But the ecological process of degradation does not occur in a political and economic vacuum. Clearing and weeding *kra wo ni* and associated species is the major agricultural task in the region. Households that can seize and control labor for the task can maintain production and capitalize on opening markets. Those that cannot will be quickly overrun by the shrubby invader and, coupled with tighter margins in non-subsidized agricultural production, be driven to increasingly marginal positions in the regional economy.

But since even wealthy families and households increasingly depend on marginal community members (poor people) for weeding labor, which they must by necessity remunerate at as low a level as possible, they fall victim to the semi-organized resistance of rural workers. These workers exercise opposition to exploitation in the only way available: working slowly, stealing crops, and otherwise contributing to lower productivity. Differential empowerment of households, in terms not only of economic power, but also of family authority, gender, and other traditional social institutions, leads to struggles over labor in the face of significant ecological change. Political ecological analysis demonstrates, in this way, that marginalization, resistance, and the ecology of species invasions are cyclically intertwined (Figure 7.1).

There are several practical lessons. First, the increasing openness of markets does little to allay the chief barriers to agrarian productivity and indeed probably only exacerbates the issues of rural labor dependence and tight margins. Only a thorough effort to overcome the structures of economic stratification in the region, perhaps coupled with subsidized control of *Chromolaena odorata*, can break the cycle of marginalization and land degradation. In a period when the state is retreating from agriculture in Ghana, solutions probably lie in state intervention.

There are processes at work, however, that typify more general assertions in political ecological analysis. Here, as elsewhere in the world, economic and political change predicates ecological transformation. In turn, marginal households become less able to secure the labor or capital inputs to manage changing ecosystem conditions,

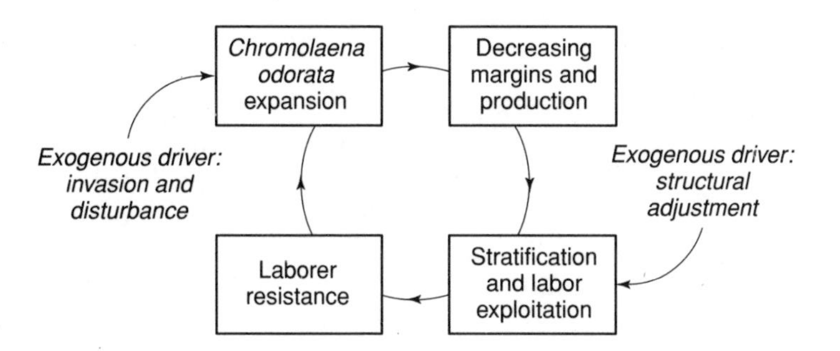

Figure 7.1 Invasive species and labor resistance in rural Ghana (Based on Awanyo 2001)

leading to more degradation. The cyclical positive feedbacks in this pattern flow between increasingly marginal communities and decreasing productivity of the land. In addition to its own fascinating, context-specific elements, Awanyo's analysis fits into a larger body of political ecological theory: the *degradation and marginalization thesis*.

The Argument

The degradation and marginalization thesis: otherwise environmentally innocuous local production systems undergo transition to overexploitation of natural resources on which they depend as a response to state development intervention and/or increasing integration in regional and global markets. This may lead to increasing poverty and, cyclically, increasing overexploitation. Similarly, sustainable community management is hypothesized to become unsustainable as a result of efforts by state authorities or outside firms to enclose traditional collective property or impose new/foreign institutions. Related assertions posit that modernist development efforts to improve the production systems of local people have led contradictorily to decreased sustainability of local practice and a linked decrease in the equity of resource distribution.

The theoretical underpinnings of this argument are several, and are laid out in greater detail in Chapter 3. They revolve, however, around two central assumptions that can be quickly summarized, one regarding the reversibility of degradation and the second concerning the character of producer margins under conditions of accumulation.

Degradation and reversibility

The first assumption is that *degradation* of environmental systems, especially after passing an undefined threshold, tends to require as much or more energy and investment to restore to its former state as was expended in its initial transformation. As outlined in Chapter 5, this model of degradation is not uncontroversial and specific ecological characteristics vary greatly between systems. Even so, in many cases, owing to problems of system resilience and hysteresis, degradation can have progressive momentum and be difficult to reverse.

Accumulation and declining margins

The second assumption is that with declining economic margins, especially under increasingly competitive global trade regimes and unregulated markets, costs and risks are passed downward to individual producers, who can be predicted to extract from the ecological system to balance their losses. The result is a pattern of *appropriation and accumulation* of natural capital, transformed into currency, at locations away from the site of production. When farmers cut their way into forests in order to increase production and offset tighter prices for agricultural commodities, the lost value of the forest (in ecosystem services or biodiversity) is understood to have been extracted from its location and accumulated on distant commodity markets, like bananas or coffee. Conversely, industrial inputs for production like pesticides or fertilizer are used with increasing intensity and cost, even while yields that result from these inputs continue to fall. In a world where the Net Barter Terms of Trade – the value of third-world commodities sold relative to first-world industrial goods purchased – fell to one-quarter its 1950 value by 1994, this model seems plausible (Spraos 1983; Porter and Sheppard 1998).

The Evidence

This is not to argue that exploitation did not happen in the past or under other economic formations. Nor does it mean that traditional systems of social relations are non-exploitative or entirely equitable. Nor does it imply that all degradation is strictly a product of economic marginalization. It does suggest, however, that under conditions of increasing marginality and disruptive social change, especially where sustained economic exploitation is allowed, undesirable regional-scale ecological transformations ("degradation") tend to increase in momentum and become difficult to reverse. So too, declining environmental conditions can be expected first and foremost amongst the most marginal individuals and groups, driving increased extraction and placing greater demands on the ecosystem. The case study material supporting this assertion is extensive, though by no means without ambiguity. Two cases are instructive.

Amazonian deforestation

There is perhaps no more emblematic case of the regional political ecology of degradation than that of deforestation in the Amazon. Without question it is one of the most prominent galvanizing images of environmental change in the last half century, in part because of the unprecedented rapidity of land cover change but perhaps more because the Amazon's historic metaphorical value gives it great currency as an ecological emblem. As Hecht and Cockburn (1989) phrase it in the beginning of their classic work on the question, *Fate of the Forest*, "what imbues the Amazon with such passion is the symbolic content of the dreams that it ignites" (p. 1). As a result, this area of tropical forest has received the attention of every possible environmental and political community, and has become so highly contested that fundamental and immediate issues have often been obscured.

The region is steeped in misunderstanding for a number of reasons. First, the fantasies and romantic conceptual landscapes that European colonists brought with them to the Amazon, associating the unbroken tree cover with a pristine Eden, have become a basic part of colonization and settlement in the region, obfuscating the actual ecological processes of the forest and the practices of subsistence communities dwelling within (Slater 1996; Cohen 1999; Sluyter 1999). Second, careful controls over information historically exerted by colonial authority have allowed little good information out of the region for long periods, heightening an aura of mystery (Hecht and Cockburn 1989). And in the contemporary period, even when the crisis of the Amazon has been recognized as acute, misunderstanding has followed from a tendency of popular accounts to underestimate the long-term and large-scale effects of human impact, while overestimating the short-term ones. This has meant a focus on the cutting of trees at the local scale with insufficient examination of structural forces involving events and players in other places (Vandermeer and Perfecto 1995).

Despite this cloud of mystery, there is clear evidence of significant recent changes, even though the rate, extent, and reversibility of change is hotly debated. The period since 1975 has seen accelerating conversion of tropical forest canopy to grassland, fields, and secondary forest succession, with some 10.5 percent of the "originally forested portion" of the Brazilian Amazon deforested by 1991 (Moran 1993; Parayil and Tong 1998, p. 63). The losses are by no means ecologically trivial, moreover, and the uncharted diversity of the forest is clearly at risk.

Figure 7.2 Amazonian cleared forest under grazing (Courtesy of Brad Jokisch)

Given the inherent monetary and non-monetary value of biodiversity in the forest, the romance of the forest and its Edenic associations, and the size and extent of the transformation, this change in land cover was unsurprisingly followed by an avalanche of academic and popular analyses. These attempted to heap blame for the transformation at the doors of the ignorance of poor farmers, the rapacity of cattle barons, the power of corrupt politicians, and the recklessness of multinational fast-food chains, most notably McDonalds.

Apolitical Malthusian arguments in the Amazon, which direct attention to over-population of the region and the inevitable "tragedy of the commons" that follows individual goal-seeking at the expense of the collective good, hold that growing populations in the region account for forest cutting and large-scale reckless land clearance through burning. The chaotic and poorly defined structure of property rights, as this line of thinking goes, is coupled with growing local human demands for land to create a free-for-all where a "global heritage" is squandered by poor hungry masses. This model is held popularly but is reflected rather less in academic studies. Nevertheless, population pressure continues to be a favored explanation in some studies of this region and similar forests in the tropical Americas (Sambrook et al. 1999).

In a sense, of course, this is an obvious relationship, but certainly not in the way intended by crude demographic arguments. As Vandermeer and Perfecto argue for Central American (as opposed to Amazonian) tropical forests: "such an observation is trivially true. There undoubtedly is an overpopulation of banana companies, an overpopulation of former banana workers looking for land, an overpopulation of adventurers seeking their fortunes in a new frontier zone, an overpopulation of greedy people and institutions, and even an overpopulation of ecotourists from Europe and the United States" (Vandermeer and Perfecto 1995, p. 12).

While political ecological explanation in the Amazon does not focus on banana commodity production as it does in Central America, it shares a similar drive to uncover the underlying causes of the problem and rejects traditional Malthusian explanation, pursuing instead those forces conceptually and geographically far from the site of tree-cutting, upwards along the chain of explanation from the local to the global. In the process, a political ecology of the Amazon stresses the context within which tree-cutting occurs and the relationship between the disempowerment of marginal communities and the loss of tree cover.

Some of the earliest and most forceful political ecological analyses of the Amazon that followed this line of explanation were also some of the first to lay claim to the moniker of "political ecology." Schmink and Wood's "'Political Ecology' of Amazonia" (Schmink and Wood 1987) and their later *Contested Frontiers in Amazonia* (Schmink and Wood 1992) both confronted the socio-political system of Amazonian deforestation and addressed the problem in terms of surplus accumulation. Using an explicitly materialist interpretation that drew attention away from individual tree-cutters and towards struggles for control of the forest between powerful groups, Schmink and Wood argued that as class stratification increases under conditions of market expansion, an increasingly hierarchic arrangement of groups will struggle over the "surplus" that comes out of the forest, inevitably overextracting.

Indigenous people and non-indigenous riverside dwellers and rubber-tappers in the Amazon, they insisted, were governed by communal land systems and redistributive economies that did little to disrupt ecosystem stability. As indigenous groups and small producers are drawn into market economies, which tend to organize the flow of capital into the hands of investors, landowners, and non-residents, people who have resided in the forest for generations with techniques of sustainable production are pushed aside by settlers and land buyers. Those who remain, joined by impoverished frontier peasant settlers pushed from the exterior, become increasingly linked into market production and competition with other producers. Fundamental imbalances in landholdings emerge along with falling commodity prices, causing the most marginal producers on the smallest landholdings to "overcrop" and cut forest disproportionately. Credit systems, middlemen, and commercialization of agriculture further reduce household margins, resulting in yet more intensive clearance. Where the state is involved, it serves the interests of elites, opening land settlement on the territories of indigenous communities, and, especially, encouraging land clearance for pasture (Schmink and Wood 1987, 1992). Degradation follows this process of enclosure and modernization, in turn driving more intense extraction: classic political ecology.

The practical solutions to the problem that emerge from such an analysis, Schmink and Wood suggest, require the establishment of appropriate "carrying capacity," well defined extraction limits, and the organization of class-sensitive policies to enforce them, attending to the impacts on marginal groups. Incentives for export cattle markets should be phased out, and the pattern of external market-oriented accumulation rejected.

In a similar vein, Hecht and Cockburn's *Fate of the Forest* also turns attention away from stories of rampant deforestation by the poor and explains the ecological problem to be fundamentally one of justice. Their research expands upon, but also challenges, the narratives of structural political ecology offered by Schmink, Wood, and others. Certainly patterns of control and accumulation are significant, they argue, but the notion that "global capitalism" or distant hamburger consumers are the driving forces in Amazonian forest decline is oversimple. In particular, their account rejects the notion that export markets for cattle and other agricultural commodities are driving degradation. In 1990 only about 15 percent of Brazil's beef was exported, for example, and logging was centered most heavily outside of Brazil's Amazon. Moreover, debt, a common political ecological explanation, also has a tenuous connection to deforestation; many forests were cleared long before Brazil accumulated debt. So too, subsidized credit for cattle had only a limited effect in the region; roughly 10 percent of Amazon holdings required credit.

Instead, Hecht and Cockburn focus on the geopolitical strategies of the Brazilian state, directed by the ruling military elite, joined to a local entrepreneurial class, which rapaciously enclosed the Amazon in an effort not only to promote economic growth, but to control an unruly and revolutionary populace, providing land and labor in a series of enormous projects and land settlements. With each fitful enclosure, instability over property increased while returns from investment decreased, prompting episodes of land clearance. This process was less a series of economic decisions by atomized peasant producers than a protracted and ongoing war against indigenous communities, impoverished placer miners, petty extractors, and rubber-tappers, whose expulsion from the forest was prerequisite to control. The conflict was propelled by land speculation in a highly unstable national economy, encouraging

Box 7.1 Down to brass tacks in Hecht and Cockburn's *Fate of the Forest*

Political ecologists don't write very well. For some reason, people at home in colonial forestry archives or East African millet fields can't seem to master simple phrasing and good anecdotes, nor convey to a broader audience the importance of the problems that they take so seriously. The implication of this failing is that political ecologists are not widely read and when they are read, they are not well understood.

Not so Susanna Hecht and Alexander Cockburn. *The Fate of the Forest* (Hecht and Cockburn 1989) is good political ecology but it doesn't read like political ecology; it is swift, organized, and urgent prose.

The book's project is a familiar one – to dispense with apolitical explanations of Amazonian deforestation and implicate instead political economy and the complex, coercive, and violent history of the region. Along the way, Hecht and Cockburn skewer the myths of Amazonian deforestation: that it is driven by logging and North American hamburger consumption. Rather, they demonstrate the violent subjugation of the region by local elites (first colonizers, later autocratic military leaders) chasing rubber receipts and minerals, resulting in rapacious ecological destruction and the mass murder of indigenous people. The pervasive feeling of the book is one of violence, therefore; even the photographic plates, which, besides showing farms and denuded vegetation typical of political ecology, also show generals and mass graves. The Kayapo Indians and local rubber-tapping producers of the forest become heroic in the account, not simply for saving forests, but for fighting against an industrial/military machine that has historically held a monopoly of force and shown a willingness to use it.

All this frank and direct writing should be unsurprising considering the authors. Cockburn, a long-time journalist and well-known radical writer, is now coeditor of the internationally read muckraking magazine *CounterPunch*. While Hecht's research history resembles traditional cultural and political ecology more closely – trained at Berkeley in geography, her own publications on ethnopedology and pasture dynamics were pioneering "takes" on deforestation and alternative management strategies – her work also embraces Latin American novelists. Both are trained to write.

Many of the details in this book have become dated. Logging has in fact become a more important part of Amazonian land use and land cover change since the book was written. So too, complex patterns of forest regrowth in some areas are an important ongoing dynamic that has only become evident in the last few years. Second- and third-wave migration into the frontier is also changing the nature of land cover transformation. But the power relations that Hecht and Cockburn lay bare remain as persistent as ever.

This book should be required reading for any researcher attempting to communicate their findings. By putting the record straight, by conveying human and ecological tragedy in plain terms, and by showing the baldly violent nature of ecological struggle, *The Fate of the Forest* is a model. The question remains. Why can't the rest of us write this well?

land clearance to establish hegemony (Hecht and Cockburn 1989). Marginalization is again central to the explanation, but in a very different way: deforestation is the outcome of state-directed class war.

These themes and counter-arguments continue to be pursued in more recent research on the question, with attention given either to specific land, labor, or market dynamics that drive tree-cutting at the local scale, echoing Schmink and Wood, or to state strategies for hegemony, following Hecht and Cockburn. In examining the pressures on household-level tree-cutting, for example, recent research on cattle-related impacts has demonstrated the disproportionate share of forest clearance attributable to wealthy, non-resident, elite, large landholders (Walker et al. 2000). This analysis reveals the micro political ecology of land clearance with less attention to large-scale contextual forces at work in the region.

Concomitantly, meso- and macro-scale analysis of the question demonstrates the way such pasture-driven land clearance has given way to logging-driven change. With more state emphasis on industrial development and the contracting of increasingly lucrative commercial logging agreements in recent years, commodity production and extraction play an increasingly important role in deforestation. Large markets for tropical hardwoods, state tax holidays, and generous licensing practices have paved the way for an exponential growth of sawmills through the 1990s. While marginal farmers have continued to procure land in the region, these communities act as stressors rather than forces, settling in the wake of an increasingly mechanized forest extraction regime enabled by the state (Parayil and Tong 1998).

While this research powerfully substantiates rejection of apolitical approaches to the issue, it is not without problems. Specifically, degradation and marginalization research in the Amazon, even in its recent, sophisticated, and multi-causal form, depends upon a model of ecology that is increasingly being called into question. This traditional political ecology stresses the catastrophic and permanent character of the crisis, and the need to identify "the root causes of this *irreversible* environmental change" (Parayil and Tong 1998, p. 63, my emphasis). This model is challenged by recent understandings of soil and secondary succession dynamics in highly variable edaphic (soil) conditions.

Over the last few years, regrowth of initially deforested areas has been observed in many areas. While ecologists have historically emphasized the acidic nature of rainforest soils and the thinness of the soil horizon – conditions that lead to leaching where nutrients crucial to forest regrowth are lost – recent field studies have shown that some areas are suitable for secondary succession of forests much like those lost in initial cutting. In particular, the Alfisols of many Amazonian sub-regions can sustain clearance and abandonment, with rainforest recovery following disturbance, depending on the land use in the period after clearance (Lu et al. 2002). Cyclical patterns of cutting, settlement, and abandonment suggest that current rates of deforestation may not be permanent and that in many areas canopy will be restored.

This merits some reconsideration of the "crisis" narrative of Amazonian political ecology, since extraction may not lead inevitably to the permanent degradation of the land and emmiseration of its residents. In even the most thoughtful research it is necessary to avoid portraying an undifferentiated ecological landscape; in fact a patchwork of soils, cycles, and forest structures predominate. So too, the romantic notion of low-impact, steady-state extraction by pre-capitalist subsistence communities must be avoided, depending as it does on a notion of "pristinity" that echoes the Edenic vision of apolitical ecology (Slater 1996).

Having said this, the general processes and long-term nature of Amazonian deforestation are fairly well established. Regrowth of disturbed forest is extremely difficult, especially on the dominant Ultisols and Oxisols of the region. Traditional extraction regimes, rubber-tapping, and swidden systems are considerably less demanding on succession than intensive grazing regimes. The relationship between the marginalization of indigenous Amazonian communities and the destruction of forest is by no means a simple one, but it is a pressing issue.

Contract agriculture in the Caribbean

The rise of increasingly contractualized agricultural production systems in global food and cash crop trade provides another extremely compelling test for the degradation/marginalization thesis. In general terms, the increasing contractualization of agricultural sales, where a grower makes a crop agreement in advance with processor, buyer, or exporter, involves a surrender by the local producer of some measure of power and resources to the larger firm. Minimally this may assure some a measure of security to a producer in a turbulent global market. It also means, however, increasing concentration of capital and power at a higher level in the agro-food chain, turning farmer owner–operators into something akin to wage laborers (Pred and Watts 1992).

The pressures that such contractualization places on producers to grow specific kinds of products in specific quantities can be predicted to give rise to marginalization, since the grower loses control of labor-time allocation and autonomy. It might also be predicted to lead to land degradation, since the intensity of cropping and inputs are set by off-farm interests with little direct knowledge of farm-level conditions. Its general coincidence with the rise of cash crops over food crops might also be predicted to lead to food scarcity in exporting countries. These conclusions are reinforced by national-scale research around the world (Goodman and Redclift 1991).

These conclusions are further supported by more detailed political ecological analysis at farm level. Working in the highlands of New Guinea, political ecologist Larry Grossman demonstrated the way in which increasing cash-oriented agricultural activities in the 1970s, including cattle raising and coffee growing, led to declines in food security with baneful results during market busts in the 1980s. With increasing integration into highly variable regional and global markets, subsistence risks increase, especially for more marginal households, whose marginality itself developed from the economic stratification following the growth of a cash economy (Grossman 1984).

Yet Grossman's later research work on banana production in the Windward Isles of the Caribbean demonstrates the subtleties and complexities of political ecology and agricultural production, and shows that global generalizations can prove misleading. The case of eastern Caribbean bananas might appear, at least at first peel, to be open and shut. The central cash crop export of these islands is the banana, an industry originally introduced by British colonizers to islands like St Vincent, where Grossman performed his research in the late 1980s and early 1990s. The St Vincent

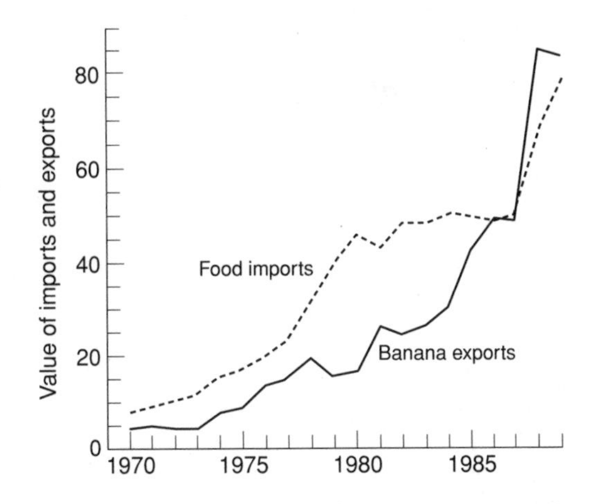

Figure 7.3 Value of food imports and banana exports in EC$ millions, St Vincent, 1970–89 (Reproduced from Grossman, figure 2)

Banana Growers Association, a statutory corporation designed by the government to help purchase chemical inputs and facilitate marketing, mostly to buyers from the UK, further supports the crop. This support and integration of individual producers, who contract delivery of banana harvests, has had a large effect on the regional economy. The value of banana exports from the island rose from 10 million to 80 million Eastern Caribbean dollars (currently 1 US$ = 2.657 EC$) over the period between 1970 and 1990.

The pieces are all in place to predict several detrimental political ecological effects, including the proliferation of chemical input hazards, the marginalization of smaller producers, and the stratification of rural communities. This is also suggested by the decrease in food crops at the expense of banana crop expansion; while banana exports grew over the period, food imports expanded in exact parallel (Figure 7.3), implying the displacement of food by bananas.

Despite the fact that food imports and banana exports on St Vincent are directly parallel, however, Grossman concludes after intensive fieldwork that there is no strong evidence to support the idea that cash crops are directly displacing food crops or leading to an increase in environmental degradation in the form of erosion and pesticide misuse. Bananas do not interfere with food crop labor demands; labor is in short supply, but is increasingly lost to time for education and, in addition, most laborers prefer to work in less demanding banana-related labor activities than in land preparation associated with food crops. Nor do these cash crops significantly displace land for food crops; intercropping of food crops and bananas is typical and indeed improves the growth of bananas. This further increases the efficiency of chemical input usage, since field preparations for food crops tend to discourage runoff more effectively than monocultural banana field preparation. This is especially true of the most marginal holders of the smallest farmsteads, those whose steeply sloping fields must be intercropped with food crops and bananas, thus encouraging the production of local foods and reducing excess chemical input inefficiencies. The poorest households are seriously affected by overall labor shortages, but these

Box 7.2 Seeking balance in Grossman's *The Political Ecology of Bananas*

Larry Grossman's research on bananas in St Vincent in the eastern Caribbean is unusual because it has a happy ending, at least sort of. Somehow, in their integration with international banana markets through contract farming, Vincentian farmers are able to provide food for themselves while increasing cash from crop sales.

This stands in marked contrast to the results from Grossman's previous work in the highlands of Papua New Guinea in the 1970s. There, examining the impacts of the introduction of coffee production and cattle raising on a village community, he demonstrated the decline of subsistence resources in the wake of market integration. People's food security tended to deteriorate with the advent of new cash crop markets, despite the promise of plenty that free-market advocates suggest.

But whereas villagers in the highlands of New Guinea were producing for "open markets," peasant banana farmers on St Vincent were working under a system of contract farming, mediated through a powerful state producers' collective. The case is interesting because it raises the issue (and the possibility) of positive state intervention into peasant production, allowing for more humane and sustainable outcomes, as well as addressing the issue of how individual households creatively balance their labor to make a sustainable living. This allows political ecology to explain not only why some systems fail, but also why some succeed.

He noted in 2002 that his research "was not intended to generalize . . . that the impacts of global capitalism on local communities in developing countries are positive." Rather, the outcomes result from "specific environmental, historical, and cultural circumstances, the nature of the Vincentian state and its welfare tradition, as well as a protected market in the United Kingdom and periodic infusions of British aid."

In writing about farmer strategies and adaptations, Grossman also sought a more "balanced" form of explanation, between highly local social and environmental details on one hand and more broad-scale political economic driving forces on the other. As he explained:

> Political-ecological studies [tend] to be sophisticated in the analysis of the political-economic dimension, but weak in relation to the analysis of what cultural ecologists emphasized – the details of human–environment interactions and patterns of resource use. The environment is more than a malleable entity molded by human activity; rather, it has significance . . . My argument was not that the environment is more important in our explanations than political economy, but that the former has not received the attention that is warranted in political ecology. (L. Grossman, personal communication, 2002)

In not assuming that social/environmental disaster is a foregone conclusion in political ecology, and by balancing local environmental details against larger policies and markets, Grossman shows a way to make political ecology relevant and useful.

equally hinder food and banana cash crop production. The extension of banana markets and their encouragement in the Windward Islands is neither making the poor any poorer, or the land less productive, at least not in the near-term (Grossman 1993).

Many of the exceptional political and ecological characteristics of the Windward Islands are notable, however. The system of hillslope intercropping is unusual in the region, and indeed somewhat unusual on St Vincent, occurring mostly on its southern end, the site of Grossman's research. The specific strategies of the state are also unusual; the use of banana packing technology and organization, subsidized by the state's statutory corporation, discourages the common problem of crop theft by removing harvested products directly from the field and so allows more stable profits from bananas than food crops. This too is unique to the region. Perhaps most importantly, the complex relationship between the Windwards, its former colonizers the British, and Geest industries, a transnational banana firm, has resulted in relatively secure and stable marketing conditions from the grower's point of view. Recent World Trade Organization rulings against the European Union's preferential licensing for banana importers may annihilate these gains.

The case of eastern Caribbean contract farming therefore shows that marginalization and degradation are by no means necessarily and absolutely linked, and that conjunctural forces, especially state policies and trade conditions, are crucial to understanding specific political ecological outcomes. More to the point, it demonstrates that political ecological analysis need not only demonstrate degradation, but can also explain the absence of degradation, especially where industries that are encouraged and protected by the state can engage with the global market on terms more equitable to producers.

Evaluating the Thesis

There are several challenges facing this ambitious thesis. The impacts of "traditional" pre-capitalist forms of land use, for example, raise several questions. Clearly land clearance in traditional subsistence production can have significant impacts on land cover, even in the absence of fully capitalized and market-integrated production. New research on human impact in Amazonian forests, for example, seriously calls into question Schmink and Wood's bold assertion that "subsistence societies, whose activities have only a minimal impact on the natural environment, approximate steady-state economies" (Schmink and Wood 1987, p. 41). So too, an underspecification of what constitutes degradation, whether it be loss of diversity, loss of productivity, loss of usefulness, etc., leads to overgeneralized evaluations of both contemporary change and the metabolism of earlier societies.

The serious introduction and consideration of regional environmental variation and variable state policy also make universal assertions problematic. Grossman's analysis of contract farming in the Windward Islands is a good illustration of this. All the conditions for predicted marginalization/degradation linkage appear to be in place: colonial and postcolonial cash crop introduction, state-sponsored input subsidies, falling food production statistics, increasingly contractualized farming systems. Yet on careful analysis, the system shows relatively sustainable and equitable outcomes (relative to many other cash crop-producing areas, in any case).

To sum up, there is a logical as well as empirical relationship connecting social processes of declining income, reduced landholding, and decreased security to ecological processes of species invasion, soil fertility decline, and forest biodiversity loss, even while simple and linear relationships are somewhat elusive. Amazonian forests are cut by marginalized settlers under market constriction, but with varying rates and trajectories of possible regrowth. Contract farmers do replace food crops with cash crops and utilize high-cost and soil-exhausting high-energy inputs, but only where the state and parastatal organizations poorly negotiate the terms of collective bargaining.

This raises some further questions about generalization in political ecology. Not all cases fit neatly into the simple pattern of degradation and marginalization. Careful examination of the specific policy environment and ecological conditions of production seriously complicates the general model as well. Even so, as Grossman points out, the lack of a degradation/marginalization cycle in St Vincent by no means rules out its prevalence globally and its existence on a case-by-case basis, as his previous research on similar problems in New Guinea demonstrates (Grossman 1984). Does this condemn political ecology to a role as a descriptive idiosyncratic science? Does case study analysis only illuminate exceptions, rather than rules?

Clearly not. What research in the field has revealed are the *processes* and operating *influences* that link degradation to marginalization, while demonstrating the configurations (in terms of ecology and power relations) under which these linkages are most likely, including conditions of available labor, crop diversity, land markets, and non-domestic species growth. The absence of degradation and marginalization patterns in St Vincent, for example, was specifically the product of state politics, local production patterns, and economic relationships with former colonial powers.

The question is not, therefore, to "prove" that degradation and marginalization coincide; they commonly do, and with effects that have a profound influence on *who gets to eat* and who does not, *who is forced to migrate* and who is not, and *who controls the labor of others* and who does not. Rather, the degradation/marginalization thesis is less a "generalizable" theory of some kind than an analytical framework in which to approach a problem. Cases where degradation and marginalization do not occur require explanation in the same way. By approaching the problem from a perspective that connects markets to hillslopes and trade policy to intercropping, it is indeed possible to make prescriptive claims about the conditions that make market integration sustainable and that do not lead to degradation and marginalization, specifically where there are strong collective contracting arrangements and state subsidy.

Even where the jury remains out on poverty and environment relationships, therefore, it is only through more detailed, ecologically rigorous, and politically contextual analysis that answers will be found. More comparative work, therefore, will yield insights, but also there is a need for the continued expansion of individual case research. The work has really only just begun.

Research Example: Common Property Disorders in Rajasthan

Operationalizing research in this area is by no means a simple matter, however. Methods vary dramatically, as do the social, economic, and environmental conditions under which research is done. I offer the following example from my own work to demonstrate the possibilities and pitfalls in marginalization/degradation research.

The research focuses on pasture and forest management in Rajasthan, India, a semi-arid state in the northwestern part of the country, where large areas of land have not been settled for agricultural development, and grasslands and savanna scrub are crucial parts of agrarian production, providing inputs into a pastoral sector that is enjoying unprecedented growth. Many experts and state officials, however, suggest that pasturage and forest resources are being destroyed in a free-for-all tragedy of the commons and that only state intervention and enclosure can preserve the lost productivity of this desertlike region. They draw attention specifically to overpopulation, chaotic and selfish producer behavior, and poorly organized practices, which need regulation by the state.

The degradation and marginalization thesis in political ecology would point analysis in an altogether different direction, however. One would be forced to ask: (1) what rules continue to exist to manage these systems? (2) are they changing? (3) are these changes and failures a product of increasingly impoverished producers overextracting to offset losses and tighter margins or are they related to cultural transformations in perceptions of authority, or both? (4) what differences do management, enclosure, or other rules systems make? Do they really matter ecologically? Is there evidence of degradation?

The work was conducted during the winter and dry season months of 1993 and 1994 in 28 villages in the western part of the state. Three of those villages were singled out for more intensive histories and interviews, while 34 sites within those three villages were analyzed to assess variation in ground cover and tree frequency, testing to determine whether people's responses to authority influenced their behavior and whether, in turn, their behavior influenced ecological conditions.

The specific research methods required for this work were as follows:

- elicit the range of accepted rule systems and their variations in village geography
- record the actual ecological practices of people across these variations and determine the reasons and motivations for differential practices (cutting, grazing, etc.) and adherence to rules
- test to determine whether ecological variations followed from behavioral/institutional variations

Each of these research tasks could easily fill its own methodological volume, but this account will be limited to just some considerations, successes, and failures of each in order to reflect the complexity of doing this kind of work. In discussing these, I hope to show that the concepts of marginalization and degradation do submit to empirical analysis, but not in any straightforward way.

Eliciting rules of use

The first task, determining the rules of use in the region's villages and their spatial patterning, was a terrific challenge. This is because many of the residents of the region had different knowledge and memory of the landscape and differing views on the legitimate systems of rules – what you can and cannot do on different parts of the landscape.

The problem was compounded by issues of caste, an important part of social reality in rural Rajasthan. As it was, I entered the field initially with a fellow researcher from a local university who was himself a member of a traditionally marginal caste: the *meghwal*. Historically a leather-working caste, they are today one of many smallholding peasant communities of the desert. Many doors opened for us as we traveled together. People, especially those from the poorest families, were eager and interested to relate their accounts of traditional and modern systems of management.

Research proceeded by listing the various kinds of land management rule systems that prevailed in each village. This was rarely a straightforward matter, and asking direct questions like "What rules are there here? Where do they spatially begin and end?" is simply absurd, nor does it make sense in the local *marwari* dialect. Instead, we used a mapping technique in which we asked what behaviors were allowed in various areas and what would happen if someone cut trees, grazed animals, or collected grasses from one area or another. We recorded almost a dozen different forms of rules, all varying not only in their form but also in the differing sources of authority behind each, and roughly mapped their outlines in several village clusters.

This was supplemented by formal surveys that queried the available assets of households, the major uses household members made of community lands, and the variation in community land demands over the course of the year. This last query was of particular importance since desert ecosystems are highly variable in productivity from season to season, and household strategies tended to vary from month to month. Rules also varied greatly, and many rules that were important during the rainy season, banning grazing on fallow lands, for example, were irrelevant during the dry season.

Silent conflicts and distrusts emerged between elite high-caste communities and ourselves, however, which often waylaid the work. Not only was my research assistant unable to enter some houses, the location where we slept at night also became implicated in the research, reducing the number of reliable conversations we could hold. Before more extensive interviewing could begin, I was forced to hire an additional research assistant from an elite *rajput* family and we all began to sleep in the shed housing the engine for the public tubewell, since it was a location viewed as socially neutral by all parties. The reliability of many of our earlier interviews (some two months of work) must, in retrospect, be treated as suspect.

Recording environmental practices and response to authority

The second task required us to observe and record the actual land uses that people practiced in the varying institutional areas. This meant not only long interviews but also days spent following goats, collecting fodder grasses, and otherwise engaging in the business of the subsistence environment. Much of the detail recorded in this process, including information on how grass is stacked to prevent rotting, how leaves are processed for fodder, and how to convince a goat to go in one direction rather than another, was never used in analysis but proved important to our overall understanding.

As work proceeded, it became clear that formal, written rules in some areas were consistently violated while in other areas they were definitely not. Some communities were inclined to break some rules but not others. These different responses were consistently linked to local social position, wealth, and caste connections. Rules mattered, though in ways influenced by informal institutions and power.

Even so, this incredible variation of practices presented problems for analysis, as did the obviously variable motivations of people in choosing one action over another. Some days, individuals choose not to herd animals on public land to which they were entitled, for example. Why? Perhaps because of the proximity of the land to the house of someone to whom they owe money. Perhaps because they associated the area with some bad luck in the recent past, like a sprained ankle. Given the necessarily small sample of households and the large areas involved, the difficulty of separating individual motivations from more general patterns of response was extremely difficult, and sometimes frankly impossible.

Determining ecological outcomes

To test whether any of these forces mattered in terms of land cover, we sampled 34 sites under four different institutional types, each representing a different kind of village authority reported in the village surveys and conversations – sacred groves, private land under limited collective access, village commons, and land under state forest department control. The problem of control was paramount. Sites were selected to be similar in as many respects as possible, in terms of slope, aspect, and other ecologically significant factors. Fortunately, except where large dunes dominated, such sampling issues could be taken into account. Even so, the challenge of turning a complex landscape into a clean grid for sampling proved to have some uneven results and sometimes required judgment calls in sampling design. It was necessary not only to sample ecological variation in a spatially stratified and random fashion but also to make sure the areas sampled were ones where we understood the rules of use and the actual behaviors of local producers in those areas. This limited our sample to a great degree.

We measured the intercept of herbaceous cover and specific species classes, in centimeters, along 100-meter transects at each site, laid in a random direction, from a point located by a grid overlaid on a local map, using a deck of cards. We also formed alternating quadrants along the length of the transects to record the presence or absence of specific species. This technique follows accepted practice for land cover analysis typically used in the region (Shankanarayan and Styanarayan 1964). The logistics of this operation, however, in an area used for subsistence production, were appreciable. On many occasions our plastic transect line was dragged away by grazing sheep and camels. Maintaining consistent recording required using the same small team for several jobs in relatively distant areas. The measurements were as consistent as possible under the conditions but, in retrospect, the sites were too few and too inconsistent for such study, and were I to conduct it today I would significantly increase our coverage.

Our results, which used a multi-variate regression to explain the variations in land cover we measured, point to some reasonable conclusions. We did indeed find a significant relationship between the type of rules/institutions in place (grove, fallow, enclosure) and the total cover of herbaceous species, the frequency of trees, and the coverage of desirable perennial grasses, controlling for the density of livestock in the area, the distance from the village center, and soil conditions.

What the results further suggest, though it is inadequately reflected in the data, is something that local people hold as a general truism: degradation of local resources is a simultaneous product of increasing dependency of the poorest households on dwindling common property resources, and the differential power social elites have in controlling resource access and rules of use. State interventions, in the form of forest enclosures, can be enforced to hold off degradation, especially if local people are given a stake in protecting them, but such a stake commonly takes the form of direct employment or subsidy. While some traditional systems thrive, like those regulating public access to private fallow land, other systems, like traditional sacred forests, are seriously endangered by economic and social changes.

Our work also showed that the increasing marginality of many low-caste communities also caused disaffection with traditional management and, in some cases, a limited adherence to new rule structures. The most elite families, on the other hand, were the least likely to follow traditional rules, like those against tree-cutting. These effects were further complicated by gender, since women were disproportionately likely to follow the traditional rules against tree-cutting in sacred forests but disproportionately likely to break the rules in state controlled forest areas. Degradation and marginalization are interrelated, but mediated by local power relationships between men and women, and between the rich and the increasingly poor.

What such results do not reveal, even with the weight of their accompanying statistical evidence from surveys and transects, are the vagaries of field-based work, including the local politics of questions and answers in a socially stratified context, the problem of representing varying points of view and accounting for idiosyncratic behavior, and the variability of ecological systems in semi-arid environments. These are all too infrequently reported in political ecological work. In marginalization and degradation research, which requires talking with people, measuring landscapes, and determining motivations under highly charged political conditions, such vagary and variability are inevitable; they simply demand more explicit consideration than they usually receive.

8

Conservation and Control

If environmental degradation is often associated with the marginalization of poor subsistence communities and working people, it might be logical to assume that conservation and preservation of environmental systems, resources, and landscapes is commensurate with community sustainability and the protection of livelihoods. This has proven far from true, however, even and especially where such communities are deeply implicated in environmental management and ecosystem maintenance. The case of Africa is superlative in this regard.

The plains and forests of east and southeast Africa have long been considered environmental wonders and justifiably received the attention of the world community as extraordinary and important sites of faunal diversity and complex ecosystem interactions. The annual rhythmic migration of the wildebeest across Tanzania's Serengeti and the dominant predators of the region, including lions and leopards, are globally famous. Masai Mara Game Reserve and Amboseli National Park in Kenya are two of the most heavily visited parks in the world, with rhino, elephant, and lions attracting thousands of visitors annually and hundreds of millions of dollars of tourism receipts. These gems of African ecology are heavily conserved and managed with an eye towards protecting a dwindling environmental resource.

Yet the management of these parks is fraught with conflict. Local people in the Serengeti trespass in park boundaries with livestock, hunt illegally, and steal wood. Why can't local people "get with the program"?

Apolitical ecology would direct attention to two factors, population growth at the park boundaries and the inherent tragedy that emerges from producers seeking individual good at collective costs. Greedy herders, in such an account, stand to gain by grazing animals in protected areas, for example, with costs borne by the state and the public more generally. The number of these invaders grows annually, since Kenya and Tanzania have annual population growth rates of 2.0 and 2.3 percent respectively.

Yet to approach the problem this way is flawed. First, it entirely ignores the problem from the point of view of local residents, who see the conflict in terms of lost ancestral resources and the risk that wild animals pose for human survival. Moreover, it overlooks the role of colonial authority in establishing and inventing the conservation tradition in the region. Perhaps most fundamentally, however, to approach the problem apolitically is to ignore the degree to which the traditional residents of the region have historically acted to help create the very "wilderness" that outsiders seek to preserve in their *removal*. So too, it means overlooking the way in which the aesthetics of the "wilderness" landscape, devoid of people, farms, and cattle, are entirely imposed by political authorities from outside the area. In sum, the problem may productively be seen as one of control over access, aesthetics, and landscape production – political ecology.

In East Africa, such analysis has been revealing. Roderick Neumann's detailed account of nature preservation in Africa, *Imposing Wilderness*, investigates the "national park ideal" and the political apparatus that enforces it in Tanzania. Through detailed historical analysis and village study around Arusha National Park, Neumann establishes not only the congruence of modern conservation with coercive colonial administration, but further shows the way the very idea of wilderness, as an aesthetic sensibility formed in non-tropical England and Germany, has been enforced at the expense of local livelihoods and the integrity of the ecosystem itself. Wilderness conservation has turned complex cultural-environmental landscapes of *production* into commodified landscapes of tourist *consumption*, where environment and society are artificially partitioned at the expense of social and ecological sustainability.

Neumann, moreover, shows that the conservation tradition and the actual territorial boundaries of the Arusha park itself are rooted in the colonial occupation of the region, first under German and then British administrations. Arriving in the nineteenth century, these governing authorities established coercive land control measures in the Tanzanian region of Mount Meru, extirpating indigenous land use practices by the Meru people who had herded and cultivated the area in and amongst wild animal populations for 350 years. These traditional human land use practices, coupled with seasonal rainfall patterns and the herbivory of wild species, actually gave rise to the biocomplex landscapes of the region, which would later ironically be enclosed to protect them from people (Neumann 1998).

Moving the Meru people into the mountain country, settlers were given control of the plains, even while many upslope regions were placed into forest reserves, where African settlement, herding, hunting, or collecting were forbidden, reducing traditional migrations and other land uses. These restrictions further evolved into strict rules for the protection of wild fauna, directed by aristocratic game hunters, and wildlife game wardens were given stronger police authority in the area throughout the twentieth century (Neumann 1996). Together, these evictions and enforcements drove the native populations into smaller and smaller areas of settlement, simultaneously robbing them of their rights to traditional methods and practices of subsistence. Equally importantly, by removing the Meru the colonizers had constructed an Edenic "wilderness" of their imagination, a land without people, which had actually *never existed before.*

When it formally emerged, Arusha National Park followed the outlines of previous game reserves closely, establishing a state controlled, non-human, management zone for the use of visiting tourists, in place of what had been Meru grazing commons a century before. But it was now supported by a global conservation discourse, which clung to a story wherein dwindling global commons (biodiversity and wildlife) demand protection from rampaging human activity. The resulting conflicts and acts of resistance on the part of local people, including cattle trespassing and wood collection, did not appear to the world community, therefore, as local land users attempting to re-establish control over the land or their rights to production. Instead these came to appear as greedy and irrational acts of uneducated locals poaching from the collective good. Efforts to gain access to the park for subsistence did not appear as a return to the integrated human–environment system of the nineteenth century, but instead as an invasion of people into a non-human wilderness (Neumann 1998).

More recent efforts in Tanzania have sought to reduce conflict by better redistributing the fruits of tourism development to adjacent local communities through formal state programs. These promise no return of traditional land rights in the park to the communities who historically lived in the region and who revere it for deeply symbolic and culturally important reasons. Nor do they provide any sort of challenge to the artificial division of nature and society, created in colonialism, which prevails in the commercial market of wilderness to foreign tourists. As Neumann points out, however, it is a first concession by the state to local livelihoods and it underlines the troubling fact that conservation is not only about control, resources, and receipts, but also about meaning, symbols, aesthetics, and the way we imagine nature "ought" to look (Neumann 1998). This well-crafted research and its convincing conclusions are typical of a second thread of political ecological investigation and argument, centered on conservation as control.

The Argument

The conservation and control thesis: control of resources and landscapes has been wrested from local producers or producer groups (by class, gender, or ethnicity) through the implementation of efforts to preserve "sustainability," "community," or "nature." In the process, officials and global interests seeking to preserve the "environment" have disabled local systems of livelihood, production, and socio-political organization. Related work in this area has further demonstrated that where local production practices have historically been productive and relatively benign, they have been characterized as unsustainable by state authorities or other players in the struggle to control resources.

The argument draws upon four fundamental theoretical foundations. First, it reflects a view that conservation reflects a form of hegemonic governmentality. Following Bryant (2002), the term "governmentality," borrowed from the work of Foucault (1991), defines a condition where consent of the governed is obtained through social technologies (e.g., conservation game reserves) and rule is self-imposed by individuals through methods of social institutions. These technologies and institutions enforce not only what people can do (rules), but also what goals and behaviors are considered socially desirable (norms and expectations) and what ecological outcomes are appropriate in the first place (aesthetics and ethics). Second, the argument depends on a growing understanding of traditional resource management strategies as institutional systems, where rules govern extraction without necessarily strong state intervention or individuated property rights. Third, it draws upon the notion that wilderness – as an imposed ideal and a produced material reality – is a social construct, specifically taking the form of nature without people. Fourth, the thesis reflects an increasingly prominent understanding of conservation territories, as bounded, regular, polygons, as ecologically and socially problematic, and inadequate to meet the goals of preservation either of wildlife or livelihoods.

Coercion, governmentality, and internalization of state rule

The history of conservation clearly reflects elements of coercive statecraft. In an obvious example, the paradigmatic national park, Yellowstone, was managed by the US military up until 1916 when the Parks Service was formed; contemporary Parks Service regalia and uniform are indeed designed to recall that military heritage. So too, traditional native community users of the area, including Shoshone, Crow, and Blackfeet tribes, were all placed on reservations in the period just prior to the park's establishment. In colonial contexts, like Tanzania's Mount Meru, these coercive histories are all the more evident in that they continue to engender conflicts over land claims. Territorializing conservation space and controlling surrounding communities is a central and primary goal in the history of environmental conservation.

Such state coercion is understood to extend beyond simply enforcing conservation rules, however. Rather, efforts center on extending the discretionary conservation power of the state by causing individuals and social groups to "internalize" the coercive missions of the government, creating *self-enforcing* coercion. In biodiversity management in the Philippines, for example, Bryant has shown how non-governmental organizations, which are usually celebrated as counter-movements to state control over local communities, have actually served state conservation goals at the expense of traditional communities. While empowering local groups to some extent, these NGOs have served to make certain state goals, like territorialization of protected areas and control of tribal groups, the internal goals of local opposition. Even while apparently opposing state control, therefore, the overall system of "governmentality" is extended even by NGOs who claim to represent marginal communities and dissent (Bryant 2002).

Disintegration of moral economy

This thesis also assumes that *social capital*, the relationships of trust and expectation between community members built through the investment of time and face-to-face interaction over long periods, is invested into traditional management systems stabilizing and regulating ecosystem flows and access to resources. The disruption of such systems is typically the outcome of significant state policy changes and the imposition of new conservation regimes. Such disruptions tend to lead to violation of traditional constraints on resource use, and to decreasing accountability in natural system regulation. This assertion draws heavily on work in institutional economics and common property theory (see Chapter 3), and has extensive empirical support (Ostrom 1990).

Where traditional systems of forest management, for example, depend on strong informal norms against tree-cutting on sacred lands, state-imposed conservation measures in these same lands are not only not respected in local practice, but they further serve to displace and shatter traditional restraints, leading to chaotic outcomes and reckless extraction. While this model does not match all empirical environmental management cases (Sivaramakrishnan 1998), its prevalence in conservation history makes it a valid assumption.

The constructed character of natural wilderness

The third theoretical foundation of the thesis involves a critical interrogation of the very thing that state agents seek to conserve, including and especially "natural" environments that require restoration and "wilderness" that demands protection. As discussed in Chapter 6, such concepts, which depend on Edenic notions of non-human nature, are constructions with little empirical support either in environmental history, where humans are implicated in the creation of many ecosystems long considered "natural," or in the contemporary world, where roads, people, and indirect human influences extend to the most remote areas. For global environmental conservation, however, this construct is commonly used to write human communities, especially those with longstanding residence in a region, out of the environmental history of a place, leaving it to lions, tigers, and other charismatic mega-fauna that are easier to market to tourists.

This coding of nature as Eden is rooted more specifically in the tendency to cast the political/economic periphery (Africa, tropical Asia and America, arid Australia) in the role of a "natural" world contrasted with the "ravaged" human landscapes of core areas (Europe and the United States). This means that one of the central imperatives of colonial and postcolonial governance is to protect and enclose nature "out there" in the underdeveloped world.

> Nature's eternity was symbolized in Africa, with its herds of wildlife, not in the plain artificiality of industrialized urban society in Europe. This perceptual polarization of "despoiled" Europe and "natural" Africa has held sway since the nineteenth century. Indeed, it was in the African colonies that early environmentalists were first able to lobby government to exert an influence inhibiting environmental changes they did not like, long before this was politically practicable in Europe. (Anderson and Grove 1987, p. 5)

Territorialization of conservation space

Finally, this argument works from an understanding that the territorialization and spatial bounding of conservation units into discrete, mappable units is in itself problematic (Zimmerer 2000). Supported by recent advances in landscape ecology as well as human ecology, this claim raises fundamental questions about the geography of conservation.

In ecological practice, the problem with such an approach is that the bounded spaces and territories typical of contemporary conservation (imagine big fenced squares or round polygons) poorly match the ecosystem functions and flows of diverse natural elements. The case of Kenya's Tsavo National Park is typical. Here, a large forested area, set aside for the protection of elephants in 1948, turned into a deforested plain with few elephants to be seen anywhere within 30 years. As ecologist Daniel Botkin (Botkin 1990, p. 19) observes, this is because the enclosure represents "the imposition of a political geography over an ecological geography" where regional-scale migrations of the great beasts could not be afforded within the restricted confines of the park's boundaries, making them vulnerable to periodic droughts and die-offs.

In social practice, the bounded, territorial, model of conservation is equally flawed. Most production systems are not spatially discrete and require the integration of different resources at different times. Livestock management, for example, requires the rotation and movement of animals through space, and many cropping systems depends on carefully managed spatial rotations and fallows.

This form of state practice is not unique to conservation; rather, it can be argued that it is inherent in the strategies and necessities of the modern state more generally. As James Scott insists, the grand scale over which states govern causes agencies, officials, and policy-makers to produce systems for measurement (maps, databases, typologies, concept-sets) that will render the world simple: neat categorical realities over which to govern that are, in his term, "legible." This legibility, and its inevitable reductions, tend towards low-resolution territoriality, just as we see in conservation areas, which poorly fit the dynamics of the local social/natural world (Scott 1998). Thus, bounding conservation reserves over traditional management spaces usually spells ecological trouble and opens the door for ongoing struggles over control.

The Evidence

It is important to note that the conservation and control thesis does *not* suggest opposition to the defense of ambient ecological systems, biodiversity, non-human flora and fauna, or areas of relatively low human impact. Rather, this argument emphasizes the degree to which such goals have historically failed, primarily because the instruments of conservation have disenfranchised traditional land managers and enforced the goals, desires, and benefits of elite communities who hold little or no investment in or understanding of ecosystem process, landscape, or local place.

In this way of thinking, political ecology is compatible with, and indeed prerequisite to, the goals of traditional environmentalism, including the celebration, protection, and maintenance of non-human nature. As the evidence below suggests, contemporary conservation not only drives traditional residents and users to the margins, it often fails even on its own terms, producing unsustainable results while perpetuating injustices and conflict.

New England fisheries conservation

Few other areas of conservation receive as much public attention as the world's fisheries, which are in distress in much of the world. The worldwide fish catch has increased more than five times between 1950 and 1995 while productivity in most global fisheries has declined. The crisis is a popular and emblematic target, therefore, for the popular press. State authorities, moreover, are embroiled in efforts to halt the problem. As the *bête noir* for conservation, contemporary fisheries management is usually characterized by apolitical criticisms that focus predominantly on "tragedy of the commons" logics and demographic explanations.

Tragedy theorist Garret Hardin's (Chapter 3) aquatic counterpart, Howard Scott Gordon, is credited with one of the earliest articulations of the "tragedy" argument in his 1953 theorization of fisheries, "The Economic Theory of a Common Property Resource." In this still-cited and much-discussed thesis, Gordon argues that overfishing is inevitable in fisheries because there is an absence of private, exclusive, property rights, so that competition over the resource by fishers who freely enter the seas and harvest at will must lead extraction past the brink of the capacity of the fish stock to recover. This problem, he argues, is compounded by the fact that, as a fugitive and invisible resource, fisheries are difficult to monitor, so that overextraction may proceed beyond a point of no return before fishers feel scarcity signals. Fish harvests must either be managed through some form of privately held and exchanged exclusive extraction rights, or some form of strong state limits and controls (Gordon 1954), a conclusion echoed in Hardin's later thesis.

The apparently incontrovertible logic of this bioeconomic thesis (following St Martin 2001) has established a general discourse of fisheries, with which all fishers, managers, and environmentalists must contend, in which (1) overfishing is understood to be primarily a product of fisher behavior, (2) the ocean is understood as an unenclosable space of open access, and (3) marine ecology is viewed as an isotropic environment of extraction.

The problems with such a theoretical geography are twofold. First, they take the fisheries question out of its biocomplex context, assuming "fishing effort" – the number of fishers, boats, and nets – explains alone the complex reproductive systems of marine life. Fish demographics, however, are governed by a set of complex interactions, many of which have little to do with fishermen and other predators.

Consider, for example, the Pacific salmon fisheries off the coast of Oregon and California. Here, what was once the third most productive salmon ecosystem in the United States has been reduced to historic low yields, with some local salmon species actually becoming listed under the Endangered Species Act. Overfishing, however, has little to do with declining yields. Rather, the flow of water from the Klamath River, upon which the salmon depend for spawning, has been all but halted through upstream irrigation development from the federal Klamath Irrigation Project, which diverts most of the water from the basin for farming (Campbell et al. 2001). Similar crises are apparent for many in-shore fisheries, which are affected by pollutants, coastal management, and distant land use practices far from the sea. A conservation approach propelled by a discourse of "too many fishermen and not enough fish" avoids sticky political problems (like regulating on-shore land uses such as farming and urban development) by focusing attention on working fishermen.

Second, apolitical ecologies of fisheries operate from a model of producer communities that envisions them operating in an open-access environment, free of constraints on entry, and with no rules to govern their behavior and catch. But most fisheries are by no means so anarchic, and systems of knowledge and informal institutions often restrain entrance to fishing grounds. Indeed, as has been shown in the case of Pacific fisheries, the burgeoning growth of many fisheries is in fact a result of intentional state development efforts to reduce foreign competition (Mansfield 2001).

Most academic and policy analyses continue to focus on the fishing problem as one of property, demography, insecurity, and tragedy, nevertheless. As a result, conservation initiatives seek to remove fishermen from the sea, and reallocate fish take based on enforcement and market mechanisms, including privatized marketable permits and rotating enclosures across a grid. Conservation, therefore, is an exercise in determining which groups or individuals are allowed to harvest fish stocks and in spatially bounding territories in which fishing may take place – conservation is control.

Political ecological research on New England fisheries has traced the incongruities between the social/ecological fishery systems and these sorts of imposed conservation mandates. New England fisheries seem, at least at first, an obvious case for apolitical approaches to the problem. Economically crucial and historically important fish stocks in the region, including cod, haddock, and flounder, have been on the decline for decades and the region's problems were the focus of discussion in cover stories for major news weeklies like *Time* and *Newsweek* (Lemonick and Dorfman 1994). The apparent driving force behind the crisis, anarchic and numerous fishers competing their livelihood into oblivion, has long been the dominant narrative to explain the problem.

Promulgated by the New England Fisheries Management Council (NEFMC), the regional conservation authority, recent efforts by state authorities have somewhat predictably followed the traditional model of the crisis and the "bioeconomic" model of its control. As explained by St Martin (2001), these approaches eschew traditional restrictions on fishing effort and their corresponding rules that limit net mesh and boat size, offering a different menu of controls, based on territorializing resource use. To limit access to what is viewed as an "open access" system, individually transferable quotas (ITQs) have been proposed, marketable quotas of specified quantities of harvestable fish stock that can be purchased by anyone and used

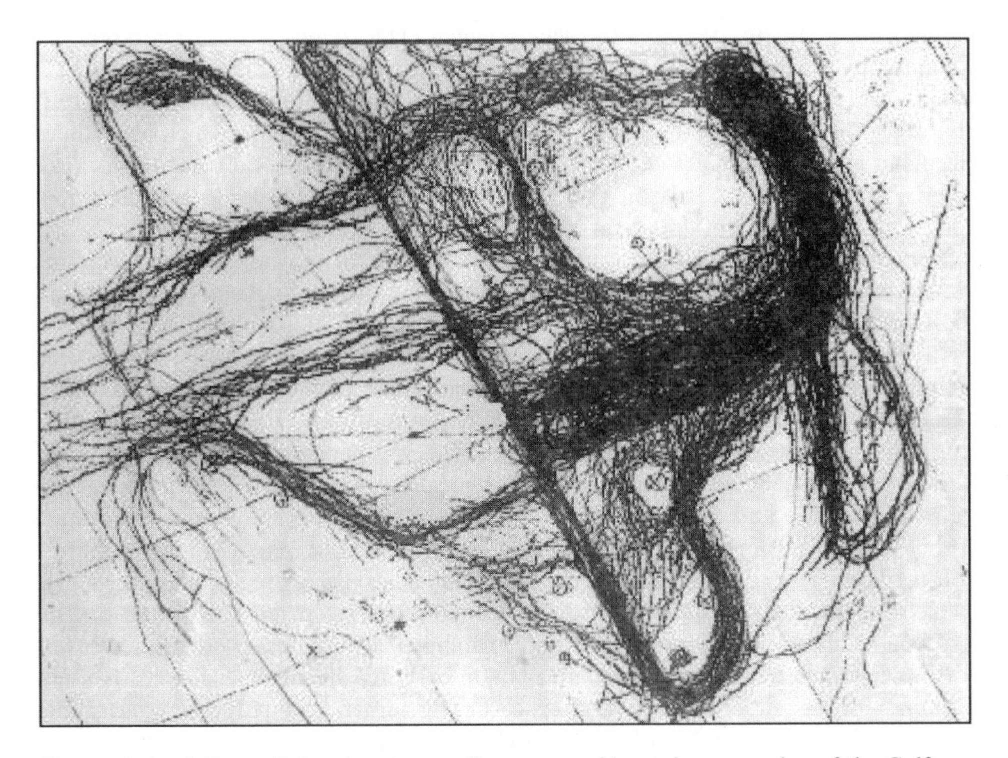

Figure 8.1 A "paper" showing the trawling routes of boats in one section of the Gulf of Maine. These represent highly spatialized and ecologically sensitive local knowledge, without which fishing in the area is nearly impossible (Reproduced from St Martin 2001, figure 6)

anywhere in the management zone. Alternatively, area management schemes have also been proposed, which direct fishing pressure to a series of management zones that open and close on a rolling basis on a grid across the Gulf of Maine.

St Martin's research reveals weaknesses in the assumptions that underlie these apolitical models and methods of conservation, as well as the social, political, economic, and ecological problems resulting from their implementation. First, based on extensive ethnographic inquiry and time spent among the region's fishermen, he demonstrates the way in which access to New England fisheries is actually highly restricted rather than fully and easily open to all. This is largely the result of the highly specialized and carefully controlled systems of spatial ecological knowledge possessed, and only rarely shared, by the region's fishermen. These systems of knowledge take the form, among other things, of "papers," carefully mapped pathways that trawling boats record and follow in harvesting fish (Figure 8.1). The high-resolution micro-geography of these maps reflects an intimate knowledge of subsurface ocean conditions and dangerous obstacles without which no boat could operate (St Martin 2001). St Martin's results follow longstanding findings in cultural ecology, which have consistently concluded that access to New England fisheries, whether for lobster or groundfish, is historically controlled and delimited by localized and traditional institutions (McCay and Acheson 1987; Acheson 1997).

Second, St Martin describes the way in which ITQs represent a serious assault on community economy and identity in the area, and pave the way for larger corporate boats, which allow for the consolidation of quotas and exclude traditional users. Drawing on related recent research that demonstrates the way in which individualized and privatized extraction leads to benefits for boat owners rather than workers, crews, or communities (Palsson and Helgason 1995), the analysis suggests the highly political implications of apolitical market solutions.

Finally and most significantly, this research points to the fundamental divisions in the spatial models of ecological process and conservation held by fishers and conservation scientists, stressing the incompatibility of bioeconomic conservation with the practice of fishing and the spatiality of marine ecosystems. Specifically, the gridlines and enclosure areas of the NEFMC follow neat, low-resolution squares, which do not follow the subsurface environmental features of breeding grounds, obstacles, and food sources so carefully traced by local fishers' "papers." The curvy and linear character of the extraction space marked in local knowledge stands in marked contrast to the boxy and territorial system of management geography imposed by state fisheries managers in US National Marine Fisheries Service maps. Management territories under official geographic organization are opened and closed in a checkerboard, which has little or nothing to do with the ecologies and use patterns of fishers or fish.

As a result, enclosures commonly force fishing effort into less productive areas or displace groups that fish in different areas into shared management areas, causing competition in reduced ranges and exacerbating rather than reducing overfishing. Different boat types and gear types (trawlers, lobster pots, gillnets) that are usually spread across diverse ecological zones are similarly consolidated with little consideration of ecological conditions. Area-management approaches like this also pay little attention to interspecies relationships and the effect of harvest quotas of one species of fish on another (St Martin 2001). Thus, political ecological analysis of New England fisheries conservation reveals a regime of tight controls that marginalizes local communities and transfers control of resources away from those who know them intimately.

What this research fails to propose, however, is an alternative to the current crude systems of conservation. The appreciable complexity of coastal zones clearly calls for a far more sophisticated method of management, and the formation of workable fisheries institutions remains imperative. The keys to making them ecologically sound, spatially sophisticated, and socially viable appear to be held in work like that of St Martin and others; explicit alternatives remain scarce, however.

Fire in Madagascar

Just as fisheries have been made emblematic symbols for commons "tragedies," fire and its intentional use as a tool for land management has unjustly come to represent the "irrationality" of traditional environmental practices. In the United States fire has long been a contested and controversial occurrence. With conservation policy in the national parks and forests services changing from suppression to "let burn" and back over the years, the politics of fire remains front-page news for many months of the year.

In underdeveloped contexts, the politics surrounding fire and conservation differs somewhat from its first-world counterpart. Where fire in the US and Canada is typically viewed as a natural hazard that poses a risk to property, in the global south it is often viewed as an irresponsible human-induced action that demands control from enlightened authorities. Fire control is commonly one of the first planks in global environmental mandates, whether for the reduction of deforestation or of carbon emissions.

The use of fire, like the use of plowing, terracing, and fertilizing, is a fundamental agrarian tool for controlling and directing environmental change. People use fire to produce and maintain pasture, to turn cut plant material into nutrient mulch, to control invasive species and insects, to clear crop waste, to aid in irrigation management, and to encourage the growth of selected species. Though a target for control and elimination by governments and environmentalists, anthropogenic fire is a building block of land management for hundreds of millions of subsistence producers around the world. The potential for conflicts around the use of fire, therefore, is enormous.

Perhaps the most prominent global historical struggle over fire is the question of shifting cultivation. Sometimes referred to as swidden or slash and burn, this category of land management activities generally refers to cultivation in regions of forest cover, where farmers clear land by cutting standing natural vegetation, burning the residue, and farming the cleared land for a few years until regrowth of forest begins, at which time the producer moves to a new forest plot (Figure 8.2).

The ecological impact of shifting cultivation has been debated. Cultural ecologists have long insisted that swidden cultivation is energy efficient and that it produces a biodiverse agrarian system structure and a physical canopy that is quite similar to the architecture of the forest it replaces (Conklin 1954; Geertz 1963; Rappaport 1975). Some work is less sanguine regarding the minimal effects of shifting cultivation (Dove 1983), but nevertheless suggests that the ecology of the system is not well understood.

Since the colonial era controls over slash and burn have been motivated by far more than environmental concerns, however. As environmental historian Richard Grove demonstrates from the case of India, among colonial managers the origins of the obsession with shifting cultivators and their agricultural practices to a large degree lay in the fact that, historically, "shifting cultivation was an inherently autonomous activity whose participants were not easily amenable to social control" (Grove 1990, p. 23). The key to controlling unruly people – tribals, peasants, and pastoralists – becomes controlling their practices, especially the use of fire. From the other side of the question, of course, individuals and groups resisting authority that they view as illegitimate or coercive have typically expressed their opposition through the use of fire (Kuhlken 1999).

The case of Madagascar is typical, both in the divisiveness of the politics over fire, and the uncertainty of its ecological impact. The island off southeast Africa's Arabian Sea coast is an impressively complex and diverse landscape. Its eastern half, dominated by tropical wet forests, is separated from its semi-arid tropical savanna western plain by steep mountainous terrain. The steep slopes of the east are covered in forests with high levels of endemism – unique native species and communities – while the long slopes of the west have fertile valleys. Roughly 2 percent of Madagascar is officially protected and the large Tsingy de Bemaraha Strict Nature Reserve became a World Heritage Site in 1990, especially for the protection of lemurs, a group of rare endemic (and charismatic) species on the island.

Figure 8.2 Swidden plots in and around Madagascar's hill forests (Courtesy of Rheyna Laney)

The forests and their biodiversity, however, are uniformly viewed as being in serious decline. The blame for this decline, moreover, is generally apportioned to the indigenous Malagasy people, the expansion of their population, and, most importantly, their use of fire. As the World Wildlife Fund succinctly insists, "the principal threats to Madagascar's biodiversity come from the small-scale but widespread clearing of forests associated with slash-and-burn agriculture" (World Wildlife Fund 2003). On the western slopes of the island, moreover, even where forest reserves are not extensive, the use of fire for pasture and crop management is viewed as a form of "ecological mayhem." The tool of fire, portrayed as a reckless practice by a burgeoning population, is the central problem according to conventional wisdom represented by state conservation officers and also global conservation groups in the Unites States, Canada, and Europe, who insist on the protection of the forest and its rare flora and fauna (Kull 2000).

This model of Madagascar's forest history and its account of fire, however, are highly problematic. First, the conclusion that deforestation is a current crisis and of anthropogenic origin is predicated on faulty assumptions about the quantity of standing forest in the pre-human settlement period. Despite claims that the whole of Madagascar was covered in forest before settlement, palynological evidence shows that the island has been a mosaic of forests, grasslands, and complex secondary succession since the last glaciation (Kull 2000). Second, the understanding of the crisis as indigenous in origin and tied to population growth depends on a faulty historical model. As Lucy Jarosz has demonstrated, most primary forest in Madagascar (some 70 percent) was harvested in the 30 years between 1895 and 1925 under colonial government supervision, with shifting cultivation joined by logging, grazing, and export crop production, especially coffee, pursued as explicit goals of the French empire (Jarosz 1993).

This raises some difficult questions about fire control in the region, especially since burning is an explicitly criminal act, as it has been since shortly after Madagascar's conquest in 1896. What are the effects of such a repressive ecological regime and its reductionist account of the role of population and fire in creating ecological decline? By placing the burden of protecting the world's lemurs and other flora and fauna on the backs of Malagasy producers, while simultaneously removing important tools they traditionally use to make ends meet, this conservation regime has created tensions between rural people and the state.

One outcome that has *not* occurred, however, is an end to the use of fire. Rather, as Kull explains, the result of these efforts has been a standoff that hinders any form of compromise: "The result of these politics – the regulation and even criminalization of a traditional agricultural practice – closes the lines of communication . . . fire is not discussed; it occurs at night, and is blamed on 'passers-by' or 'evil people.' In this context, local organization and management of fires becomes impossible" (Kull 1999).

This analysis provides a useful antidote to narratives that direct blame for conservation failures on the rural poor. It does not make the problem of fire at the boundaries of biosphere reserves go away, however. Nor does it address the complex developments in Madagascar's agricultural economy and demographics in recent years. Given that the agricultural frontier in Madagascar is now effectively closed by the creation of reserves, what kinds of producer responses are desirable and possible? With hill forests closed to local producers, will farmers respond with environmentally benign and sophisticated methods of intensification, or do structural economic barriers mean inevitable over-cropping, soil erosion, and degradation? The evidence from intensive farm-based field study remains mixed, but producer choices and techniques in the region will clearly depend on broader elements of political economy: markets for cash crops, economic liberalization, and international development pressures (Laney 2002).

Social forestry conservation in Southeast Asia

The Southeast Asian nations of Indonesia, Malaysia, and the Philippines together contain roughly 120 million hectares of tropical forest, a land area five times greater than that of the United Kingdom. These countries also suffer from deforestation, with official rates ranging from 1.0 percent to 3.5 percent per annum. As a result, they have long been the targets for conservation schemes, investment, and social/legal institutions to slow or halt deforestation.

The practices of contemporary forest conservation, however, often face violent resistance by peasant producers, forest residents, and other communities. Scarcity, population growth, and the inevitability of conflicts involving common property resources might be used to explain Southeast Asia's forestry crisis. With annual population growth rates of 1.4 percent (Indonesia), 1.9 percent (Philippines), and 2 percent (Malaysia), it might be crudely argued that demand for agricultural land and fuelwood together have served to reduce tree cover in the region.

Such a conclusion would be premature, especially when viewed through the lens of state efforts to control and manage the forest. Rather than simply discouraging local people from harvesting trees, state forestry during both the colonial and contemporary periods has been instrumental in organizing the extraction of timber, especially of valuable species like teak (*Tectona grandis*) in Indonesia and mahogany hardwoods in the Philippines. Such hardwoods covered about 10 million hectares in the Philippines in the 1950s, but were reduced to roughly 2.2 million hectares by the 1980s, owing to commercial logging contracts arranged by the state, rather than the saws of local farmers and forest dwellers (Cruz et al. 1992). Such ongoing commercial extraction, which extends few benefits to local people in terms of either payment or employment, has been coupled with a series of debt crises in the region that force the urban poor into periodic rural migrations, exacerbating tree-cutting for agriculture in already forested areas (Kummer 1992). The actions of state foresters in the region, whose efforts to aggressively enclose and harvest the region's forests have been continuously blocked by violent resistance from local populations, reveal, moreover, the way in which deforestation "crises," and the techniques of conservation they appear to require, become techniques of state control.

The consistency of colonial and contemporary forestry The case of Java in Indonesia is instructive. Surveyed by Peluso in her comprehensive critique of the state forestry in the region, *Rich Forests, Poor People*, the history of conservation is revealed to be one of struggles for political power and resistance, rather than simply tree protection and plantation.

While precolonial Javanese forests, governed by kings in complex arrangements entwined with local use rights, were by no means managed in an egalitarian or democratic manner, the entry of colonial state law and market arrangements increased the consolidation of forests under state control. As early as 1870, all non-cultivated areas were claimed by the state. Establishing forestry reserves in the late 1800s, Dutch foresters dramatically expanded their direct control of forest land so that by 1940, when the Japanese invaded and occupied Java, the area of state-controlled forest had increased to over 3 million hectares, from the 1.9 million hectares under state control in 1865 when forest laws began. Late-nineteenth-century economic relations, moreover, increasingly directed forest activities to wage labor as new tax arrangements forced increasing numbers of agriculturists to become workers in forestry extraction operations (Peluso 1992).

Equally important is the degree to which colonial forestry under the Dutch East India Company indulged in extending limited access rights to producers as a social good. In particular, the forest bureaucracy held to an ideology of paternalistic "mutual benefit" where it ironically continued to acquire lands traditionally belonging to producers while occasionally offering limited land subsidies in return.

> By the 1930s . . . the Forest Service had begun to speak already of its generosity in "lending" land to forest peasants to grow agricultural crops for a year, or sometimes two. The state assumed the role of landlord and land administrator formerly played only by local patrons and Javanese nobility. While the benefits of land and species monopoly accrued to the state, forest villagers increasingly lost legal access to forest land and species. Nevertheless, the notion that state control of forest land and forest products was for the sake of the greater good would pervade Indonesian forestry policy long after the Dutch had left. (Peluso 1992, p. 77)

Javanese producers resisted these efforts at control, at first by avoidance but later through violent and organized movements. As state control of forests persisted into the independence era and the forest service increasingly took the form of a semi-militarized police force, which collaborated heavily with large landholding elites, poorer producers increasingly plundered teak forests, harvesting trees in large communal gangs or groups.

The paternalistic approach under colonialism and the occasionally violent resistance it inspired in local populations set a political pattern so enduring that current-day forestry closely resembles that of the pre-independence government. As Peluso reveals, the consistency of the colonial and postcolonial Indonesian state in forest policy was a product of the relative stability of the forestry bureaucracy itself from one period to the next as well as of the deeply held ideologies of state control built into conservation bureaucracies.

In sum, the case of Java, like that of other Southeast Asian states, underlines the deep historical roots of conservation ideologies in forestry predicated on the removal of resource-dependent people. This ideology would continue to sit at the heart of more recent efforts to democratize forestry in the region, and so undermine efforts at contemporary social forestry initiatives.

The limits of social reform in forestry Since the 1980s reforms of coercive and paternalistic forestry have been attempted throughout Southeast Asia and elsewhere. These reforms are usually referred to as "social forestry" and include techniques that attempt to reconcile the needs of local residents with those of conservation bureaucracies. Agroforestry, social forestry, farm forestry, and community forestry are all devised to improve local economies and ecologies, simultaneously helping people adapt to changing forest conditions while working to mitigate those changes.

The limits of such reforms, however, are set in the deeply ingrained social relations of foresters and local producers, and the political policy imperatives of central governments. In Java, Peluso argues, social forestry reforms designed to increase access to forest land for locals met with very limited success, even while illegal teak poaching continued. There are several reasons for this. First, access to these programs has been limited to those with good connections to local foresters. Longstanding and deep loyalties have formed between patron families and forest managers, which limit and direct where social forestry support and resources are distributed. So too,

Box 8.1 Deep experience and Peluso's *Rich Forests, Poor People*

"Peasants," Nancy Peluso writes towards the end of *Rich Forests, Poor People*, "file few reports, write few letters, issue no legal guidelines or justifications for community and household use of the environment" (Peluso 1992, p. 252). Tracking the history of a people "without" history therefore remains a driving motivation for many political ecologists, from James Scott and Michael Watts to E. P. Thompson. It presents a methodological challenge for political ecologists, however, in attempting to reconstruct earlier eras of resistance when formulating arguments for rights of local people overlooked in development.

While this serious barrier can never be fully overcome, Peluso's text suggests that attention to people's own accounts and long-term and sustained research into a place and its politics can shine light on hidden historical processes of resource control, degradation, and recovery. Prior to even beginning graduate school, dissertation research, or conceiving her book, Peluso spent nearly six years in Indonesia, including a year as a member of a Man and the Biosphere (MAB) research team in east Kalimantan.

MAB, a United Nations program to foster sustainable development and biodiversity conservation, put researchers on the ground around the world to investigate resource use in long-term, field-based explorations. The MAB experience, combined with Peluso's three years of living in a rural Javanese village prior to her time in east Kalimantan, revealed political and intellectual contradictions for her, which she would explore in the research on which *Rich Forests* is based. While her associations with neo-Marxist anthropologists in Java drew her attention to agrarian political economy, her experience with MAB emphasized environment and human ecology; yet each group worked in isolation from the other.

Peluso's critical ethnography of political struggle was born of her urge to reconcile these experiences, her frustration with the increasingly anti-human agenda of conservation, and her interest in the consistency between two histories of oppressive exclusion in the colonial and postcolonial eras. As she told me in 2002:

> The first of these was some 150 years of exclusionary management of Java's teak forests, mostly for logging and conversion to teak plantation management, wherein people were regarded merely as occasional labor or as pests, squatters, and encroachers. The second entailed more recently developed state tactics of declaring and mapping extensive tracts of long inhabited land as uninhabited forest for the benefit of state actors and corporate interests, as was the case in Kalimantan. International conservation interests and organizations entered into this fray in the name of saving forests, but oftentimes ignored the interests of the local people.

Ironically, the deep experience and knowledge that resonates throughout *Rich Forests, Poor People* reflects, at least in part, opportunities and perspectives born of Peluso's experience of ugly politics in the world of conservation. Nor is this unusual. Many of the field experiences and personal interactions with people and environments that drive political ecological research have been accumulated while researchers pursue careers in environmental and development science. As Peluso herself notes about her book, it was "a culmination of ideas and experiences that had brewed inside my mind for several years, as well as a new beginning." Peluso's trajectory, and the urgency of her transition, represents a common and important kind of political ecological biography.

the poorest producer families, even where they can gain access to land through such programs, lack sufficient capital to effectively participate. Structural barriers have consistently foiled the most apparently far-sighted and progressive reforms (Peluso 1992).

The barriers are far more persistent and pernicious than even that, however. Comparative forestry research in the region by anthropologist Michael Dove, among others, underlines that these failures in social forestry are located in the divisions around defining the meaning of land degradation and even of forestry. The extent and causes of the problem are by no means agreed upon, and participants in the conflict are locked in deeply rooted beliefs. As Dove explains, "the first and perhaps most heatedly debated question in current social forestry discourse in Asia is the explanation of what degraded the forest resource in the first place," with foresters identifying long-term population-driven forces and locals stressing sudden exogenous events driven by officials and outsiders (Dove 1995, p. 318). Moreover, the ideology of locals, rooted in resistance and autonomy, conflicts starkly with that of foresters, who continue to view themselves as exclusive custodians of forest resources.

Finally, the norms of governance that prevail in the region, even where they are challenged by formal policy reform, remain unshaken. These "vernacular" models of development are ingrained and resilient, if unacknowledged in official policy statements, assuring a continued flow of resources from the forest periphery to the government center, despite efforts to reverse this. By this way of understanding, social forestry fails, not simply because of a few local "bad eggs" and corrupt individuals, but because it poorly fits the extractive model of the national system and the silently held ideologies of national control (Dove and Kammen 2001). At bottom then, even as social forestry and other collective means of co-management developed to cope with the failures of conservation forestry, conservation represents control, largely because the overall model of development does as well.

Evaluating the Thesis

The accumulated political ecological evidence demonstrating the political and ecological dysfunctions of environmental conservation, only briefly summarized here, is compelling. Clearly, militarized command-and-control and colonial legacies of development, backed by apolitical Malthusian and "tragedy" narratives, persist in contemporary conservation, leading to both inequity and failure.

Riven bureaucracies and efficacious species

But these bold arguments often neglect crucial complexities and opportunities inherent in conservation efforts. First, they often overlook the complex divisions and contradictions within the conservation agency itself, which lead to unexpected outcomes and possibilities for environmental reform. Second, these explanations sometimes ignore the role and effect of non-human actors in the conservation process and seldom extend demands for justice to flora and fauna, setting up false dichotomies between tigers and farmers, diversity and production.

In the first case, many critical histories of state conservation tend to characterize in somewhat monolithic terms the "ideology" of the state forest services or conservation agencies. Even comparatively rich social histories of agencies tend to describe single "institutional" cultures that determine what is normal, acceptable, and desirable. Complex internal divisions are thus reduced to limited caricatures. The state is an extremely complex entity, however, and just as local communities are riven with gender and class divisions, even small groups of similarly trained conservation professionals can differ in their imaginations and goals. So too, state conservation policy can divide and subdivide over space and time as it articulates with local community politics.

The implications for such oversimplifications are both conceptual and practical. As Sivaramakrishnan observes in his survey of colonial forestry in Bengal, the implementation of rigorous fire protection regimes came into conflict quite early with local agrarian practice, including burning for pasture and field preparation. As a result, forestry field officers, observing and struggling with local practice, soon came into conflict with ecological experts in distant offices. The occurrence of major forest fires and other political ecological events caused hardline conservation fire control to become more nuanced and differentially applied. Local and state knowledges mixed, conflicted, and produced new outcomes in an internally divided conservation bureaucracy (Sivaramakrishnan 1996, 2000). In similar work on the contemporary evolution and application of local tree knowledge in India, Brodt demonstrates the pliability and social contingency of "traditional" ecological practices (Brodt 1999). The more general implication is that normative distrust by political ecologists of state science and expertise and its artificial conceptual separation from local knowledge and practice may serve to improve neither the equity of environmental management nor its ecological effectiveness (Agrawal 1995; Rangan 1997; Dove 2000).

Secondly, non-human elements of conservation ecology are sometimes lost in conservation and control research. Generally, political ecology in this area proceeds from an anthropocentric perspective, which reduces the efficacy and practicality of the explanation in two ways. First, it underplays the role of animals, plants, and soil in delimiting and directing conservation histories. Yet these players can produce profound effects in alliance with other actors. Fast-growing plantation tree species may shelter and reproduce some elements of local ecology in conservation but drive others out, for example (Robbins 2001). Squirrels, pigs, and cats have altered ecological history as profoundly as human-caused fire and timber harvesting (Crosby 1986).

By defining the effects of conservation as the control of landscapes by specific human groups (foresters, farmers, herders, etc.) rather than groups of species (livestock, grasses, and humans), the complexity of ecological history is ignored. Equally, such an approach leaves no room for demands for justice on behalf of non-humans, whether charismatic or not. This suggests a woefully limited normative vision, which further ignores the way in which the environmental histories of both marginal people and animals – Native Americans and wolves, Indian forest dwellers and tigers, East African herders and lions – reveal the simultaneous coercive elimination of vulnerable people *and other species* (Emel 1998). Finally, such an approach suggests practical political limits, since it makes it hard to politically ally with concerned environmental groups whose sympathies may lie with tigers, elephants, or lemurs. To overcome these limits, as will be argued in Chapter 11, it may be necessary to examine and acknowledge how trees and animals form "alliances" and networks with human groups to establish and reinforce specific outcomes.

Alternative conservation?

What does this portend for understanding alternatives? What prospects exist for progressive environmental management that couples local justice with protecting desirable non-human nature, including charismatic carnivores, valuable herbaceous species, and complex ecosystems and landscapes? Can concessions to local people make conservation viable? The current record of such efforts is unpromising.

Even where some local land uses are permitted on the fringes of conservation territories, as is increasingly apparent in the new "buffer zone" approach to park management, the kinds of land uses and appropriate behaviors of local residents are placed under increasing control and scrutiny. As Nuemann (1997) explains in his survey of buffer zone approaches in Africa, the approach depends on a romantic and exotic view of residents as primitives, whose use of some conservation boundary areas is tolerated as long as they uphold a socially undifferentiated and traditional pattern of behavior, as if they were part of the fauna of the park. So too, settlement of land claims for local communities is never straightforward or unproblematic, since the processes of relocation and enclosure during the colonial era, coupled with the complex social differentiation that grows from ongoing regional development, make any simple return of land rights highly political.

Buffer zone approaches, therefore, operate from an image of traditional society existing in harmony with nature that: "precludes any analysis of social differentiation and agrarian change, or understanding of rural communities' linkages to a larger political-economy . . . This conceptualization ignores the historical forces which link underdevelopment and environmental degradation in Africa" (Nuemann 1997, p. 575).

Thus, there is no "going back" to a conservation regime whose concession to local communities is to admit "primitive" practices along a buffer. So too, any such state grants to local people, as in social forestry throughout Asia, will do little to undo the control function of conservation, since they fail to challenge the *process* whereby conservation territories are established and managed, which remains coercive and state-centered.

Comanagement and participatory efforts, though beyond the scope of this survey, offer some promise in this regard, as they appear to decentralize control over conservation from state to local authority (Jeffery and Vira 2001). The common assumptions in their implementation, however, that communities are identifiable and discrete units and that community involvement is largely a problem of defining property rights, are flawed (see also chapters 9 and 10). Joint forest management in India, for example, which extends to local communities control over forest conservation and allows profit sharing from timber sales, marks an important departure from traditional practice, but it raises the possibility of new conflicts between communities by fixing and territorializing complex systems of use rights (Sivaramakrishnan 1998). Clearly then, the lessons of political ecology, which stress the entrenched systems and coercive character of territorialized environmental control, will remain essential to alternative conservation models as they emerge.

Research Example: The Biogeography of Power in the Aravalli

Implementing conservation and control research demands multiple methods and sustained time and resources. Continuing with an example from my own research, an investigation of conservation impacts in India, I present an example of the rewards and difficulties of creating a comprehensive account of land history, land use, and land cover outcomes from state interventions, pointing in particular to the gaps, estimates, and guesswork inherent in any such project.

A classic case of conservation and control?

This research focuses on the Godwar region of Rajasthan, located to the southeast of the research site described in Chapter 7, along the spine of the Aravalli hills. These hills are the highly eroded granite remnants of Precambrian uplift and divide the humid southeast from the arid northwest of the state (Lodrick 1994). Unlike the more desert-like Marwar region of the northwest, this is an area of relatively good rainfall, high groundwater levels, and reliable aboveground runoff, which provide the resources for irrigated agriculture in both the wet and dry seasons. The hilly forest area receives around 500 mm of rainfall annually and is dominated by a range of tree species, including *Anogeissus pendula*, *Butea monosperma*, and *Ziziphus nummularia* (Jain 1992). The plains adjacent to the hills (Figure 8.3) are dominated by farms and pasture consisting mainly of grasses, especially *Cenchrus* species and *Cynodon dactylon*, and drought-tolerant trees (Robbins 2001).

In 1986, 562 square kilometers of the forested hills were enclosed to form the Kumbhalgarh Wildlife Reserve, a wildlife park managed for panther, hyena, and sloth bear species. The enclosure is one of the few locations in India where the wolf is still breeding successfully and other important species, including nilgai (*Baselaphus tragocamelus*), wild boar, and langur monkeys (*Presbytis entellus*), provide an important food base for predators (Chief Wildlife Warden Kumbhalgarh Wildlife Sanctuary 1996). Moreover, many areas on the surrounding plains have been identified as areas for social forestry development, and tree plantation is common in several other village reserve areas.

The enclosure of the forest did not come without conflict. Local producers have long used the hills to procure building materials, fodder, agricultural nutrient inputs, and thatch, while many of the poorest households draw on the forest for medicines and famine foods. The enclosure of areas from human use for wildlife protection, coupled with development of social forestry plantations, appears to be a classic case of *conservation and control* but many obvious questions arise in looking at the forest from a distance. Have local groups traditionally had access to the forest? How does forest use fit into the regional agro-ecology? Has enclosure meant a loss of household resources? Have plantations offset community use needs lost in enclosure? Have local users resisted enclosure with illegal forest extraction? What effects have enclosure and producer adaptation had on land cover?

Figure 8.3 The densely populated agrarian region surrounding the forested Kumbhalgarh Wildlife Reserve in Rajasthan, India – plain seen from the hills and hills seen from the plain

In answering these questions, I undertook two seasons of fieldwork during 1998 and 1999, surveying a wide range of household types, analyzing satellite imagery, interviewing foresters, and following producers into the forest. The work yielded several useful results but became snagged on three difficult methodological sticking points, especially problems in:

- establishing historical patterns of access
- understanding the contemporary land uses and enclosure impacts
- tracking unintended consequences

Establishing historical patterns of access

The enclosure of Kumbhalgarh was not an unprecedented one; state conservation in the area is longstanding. But the records that describe the nature and extent of conservation are scattered around the globe and often buried in the memories of aging locals and retired foresters. Initial investigation of the history of the reserve depended on current reports, review of relevant laws, and discussions with retired foresters. This revealed that the use rights of local producers were encoded into the Rajasthan Forest Act of 1953, which established the reserved forest, set fines and punishments for violation of forest land, and allotted access rights and nominal fees for all those who demonstrated *traditional use* of the forest. Traditional use was generally defined by residential proximity to the reserve, a system that has significant faults, given that many traditional forest users are migratory herders from beyond the region.

Further conversations with older foresters, over countless cups of tea and whiskey, revealed that local use was long accompanied by commercial and industrial extraction. The forest was managed under a private contracting system (*thekidar*) from 1955 to 1969 and later under a direct Forest Department marketing system (*vepar vibagh*) from 1970 through 1983. In both cases many important tree species underwent heavy extraction, including khair (*Acacia catechu*), safed dav (*Anogeissus latifolia*), and karaya (*Sterculia urnes*), three species reported to be currently scarce. These results are important, especially in evaluating any argument that attributes forest decline to local users, since intensive commercial extraction has a long history and targeted the very species that are now in decline. The establishment of the wildlife reserve, in part an extension of increasing focus in national policy on biodiversity and in part a product of international funding for wildlife habitat protection, would establish new controls, but only after a century of state-sponsored extraction.

Richer and deeper analysis of the reserve's history, especially in the colonial period, proved more difficult. Because Marwar was a semi-independent principality, unlike some adjacent kingdoms that fell directly under British hegemony, forestry management records, to the degree that they were kept in the earliest periods, are sketchy. Records on forestry in Marwar are especially thin, and a 10-day visit to the India Office records in the British Library proved useful, if over-brief. There, state reports from the late nineteenth century are kept in well-ordered volumes, indexed by date, handled in an organized fashion in a well-funded and climate-controlled collection.

These records reveal the tradition of spatially variable use rights for grazing and forest product extraction and the origin of a territorialized system of rotating blocks, in use to this day. The initial colonial-era report from the area states, for example: "The area for grazing is very large, so that closing a few blocks entirely would make no difference to the villagers. I would not stop grass cutting, which should be allowed in these blocks for two months, November and December" (State of Marwar 1887, p. 27).

The records are far from comprehensive, however, and 10 days is far too short for any serious investigation of historical management, as any environmental historian will tell you. Block numbers and descriptions of reserve areas, for example, have no matching maps from which to launch a field-based investigation of current conditions. Many gaps between years leave holes in the record of administrative change. Actual practices on the ground, the kind that might appear in the journals of field foresters, are obscured in administrative documents of this kind. In sum, many central questions of how the system of conservation at Kumbhalgarh had changed in the last century were informed by historical survey, but robust details remain difficult to secure. We are left with only a sketch.

Understanding contemporary land uses and enclosure impacts

Contemporary land uses in the region are somewhat easier to determine, but require more footwork. Research assistance was acquired by working through the Lok hit Pashu Palak Sansthan (LPPS), a non-governmental organization that represents the interests of local herding communities who depend heavily on access to the forest, and the chief forester/warden of the reserve. I have long had a good working relationship with the LPPS, for whom I have produced summary reports used in advocacy. The LPPS was interested in the survey results, as was the chief forester of the reserve, who considered his own information on local forest use to be limited. Working from these bases, and recalling previous experiences with local caste divisions, I hired two assistants, one from the warden's *rajput* family and the other from the *mali* (gardening/horticulture) community. Together we wrote a simple but comprehensive questionnaire that we used to quickly survey local producers' usage of non-domesticated species. Administering the survey to 157 individuals in eight villages bordering the reserve, we stratified the sample to be representative of local caste divisions, including 20 women and 18 foresters at varying stages of their careers and levels of bureaucracy. Most interviews were long, slow affairs, conducted while walking with herds in fallow fields, smoking cigarettes, threshing wheat, or grinding opium.

The results were revealing; the survey showed a total of 79 non-domesticated species important to household production and 113 species/uses in all, most of which were described as essential as inputs in farm and herding production. Most importantly, these wild species were not used as simple subsistence or famine-security inputs, but instead were fundamental inputs for capitalized production in the region. Moreover, they were heavily drawn from forest lands and village fallows. In the absence of offsetting inputs, conservation measures in the last few years, with decreased access to non-timber products, unsurprisingly meant decreased receipts, yields, and margins for almost all families in the region.

To confirm and understand the relationship between forest access and household inputs required many days of walking in the reserve, accompanying herders and wood, bark, and leaf collectors. It further involved an informal mapping exercise, often accomplished with a stick in the dirt, where people described the geography of their collection patterns and the location of important species. Such efforts were usually unproblematic, but on occasion inspired distrust; just as the "papers" of New England fishers represent knowledge that helps to control and delimit open access to the commons, the locations of graze, browse, and collection species are sometimes considered proprietary. Occasional violation of official enclosure boundaries, moreover, made some daily herding paths technically illegal; while most herders were unconcerned about their movements becoming public knowledge, many were less sanguine. Comprehensive maps of land use in and around the reserve, as a result, remain highly incomplete.

Tracking unintended consequences

Tracing the history of the reserve and indexing the dependence of local producers on a changing resource base provides a compelling picture of the way increasing national and global mandates for biodiversity preservation translate into costs and controls in the lives of regional producers. The ecological impacts of such conservation mandates, especially when coupled with the history of species plantation in the region, are less clear, however. To explore this final link in the political ecological chain, we overlaid two satellite images of the area, one from 1986, when the reserve was established, and the other from 1999. Both images came from the dry season and both came from years of comparable rainfall.

The later image was ground-truthed for accuracy, a slow and painstaking process. Dozens of ground points were established in locations stratified across the area of the image (some 900 square kilometers) and each spot was visited to determine land cover. Some of these spots were far from roads and involved a fair amount of legwork. These ground values were then cross-tabulated against the land cover values established in the image to determine the accuracy of classification. In our case, we tested for an accuracy of around 80 percent. This means that one out of every five pixels was misidentified, a high figure given the variability of land cover, but certainly not one unusual in remote sensing applications; accuracy values are too often unreported in publication from this sort of analysis.

When the analysis was complete we were able to produce a coherent map of land cover change over a 15-year period. Changes in the area were rapid. Besides the obvious increase of dry-season cropping and the perilous decline of pasture resources for the region's large herds, a product of state and private investment in tubewell irrigation, there is also a startling increase in the amount of tree canopy. Indeed, tree canopy cover appears to have expanded by nearly 50 percent across the region.

But, as described in Chapter 6, this canopy cover largely consists of *Prosopis juliflora*, an introduced scrub tree species of little value in either wildlife conservation or local production. There are several reasons for this expansion, all of which are the indirect outcomes of state conservation goals. On the ecological side, decreased access to pasture resource, in part as a result of forest enclosures and in part because of state-subsidized agricultural extensification, means more animals feeding on less

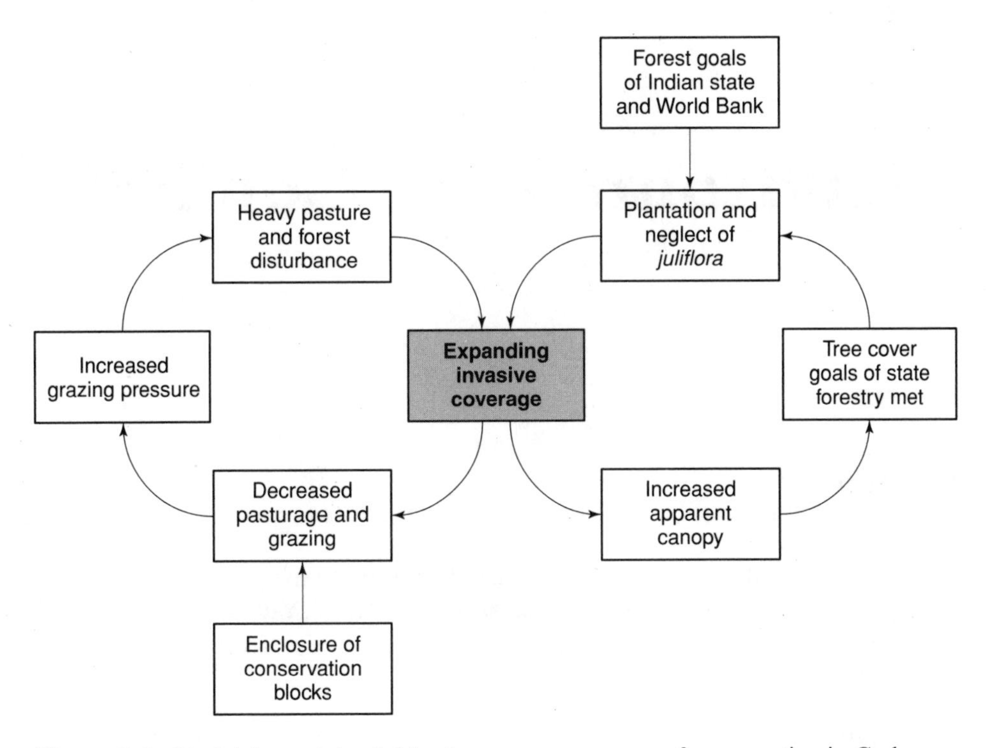

Figure 8.4 Explaining unintended land cover consequences of conservation in Godwar

available land, in both village pastures and adjacent forests. This in turn results in increasing grazing and browsing disturbance, which removes diverse ground cover, allowing invasion of these scrub trees into non-competitive environments (Figure 8.4).

On the bureaucratic side, the expansion of canopy cover fits well with the goals of the national forest policy, reinforced by recent Word Bank mandates, to increase the cover of forest in India to 33 percent. As incentives to meet regional land cover goals are high, efforts to slow or halt the expansion of the species have been minimal. Even so, many foresters themselves express concern that the rapid expansion of tree cover may be anathema to their wildlife conservation goals, drawing into question their future tolerance for this tree. The investment required to halt or reverse the spread of the species, however, may even now be beyond the fiscal and manpower resources of the state, leading to some counterintuitive conclusions. The efforts of the state to control land through conservation has not only reduced access for local people to traditional resource lands, it has created ecological outcomes increasingly out of the control not only of locals but of the state itself.

Even so, the methodological demands of explicating and confirming many of these links are high, and proved in some ways beyond my own capacity. Historical land records remain sketchy, comprehensive land use patterns remain incomplete, and land cover change dynamics remain snapshots at best. As a result, a full account of causes and effects of conservation is still elusive.

9

Environmental Conflict

Lake Nakuru, in Kenya, is a remarkable conservation regime, both in terms of its natural beauty and its apparently forward-thinking management structure. One of Kenya's premier national parks, the enclosure is home to wild game like giraffe and rhinoceros, and dawn on the lake is seasonally marked by dense flocks of pink flamingos covering the water. Management of the park, moreover, is increasingly sensitive to the artificiality of the park's own boundaries and the inevitable cross-boundary interactions between the park and the human-populated landscape beyond. Under the guidance of the World Wildlife Fund, management of Nakuru has extended to include a watershed-level management plan, specifically encouraging the adoption of land use practices and conservation on surrounding private lands in an effort to improve siltation conditions that may contribute to degradation of the lake's ecosystem. Through education and planning with local land users the state is seeking more sustainable land management outside the park to improve conditions within. A popular national park, an integrated approach to ecology, a recognition of human influences; so far, so good.

As Daniels and Bassett demonstrate (2002), however, all is not well with these conservation efforts. Adoption of land use controls is limited and sporadic. Where farmers have accepted new methods, including tree planting and terracing of fields, many such improvements have fallen into disrepair and neglect. Based on research in the region, Daniels and Bassett conclude that what should be an apolitical process of land improvement and wildlife conservation is nested in a context of instability.

Violent clashes in the rift valley of Kenya, where Nakuru is situated, have risen throughout the 1990s, marked by death and property damage. While these conflicts fall along ethnic lines – specifically between ethnic minority groups (Kalenjin, Massai, Turkanan, and Samburu peoples) and other communities (Kukiyu, Kisii, and other groups) – they are further situated in electoral politics and land control in the rift valley region. Here, minority groups, especially pastoralists, are represented by the ruling KANU party that has been pushing in recent years for a political *majimbo* system that would give these indigenous groups greater power and control over land in the region. The opposition parties, representing other groups, have in the process become targets of political protest and frequently violent attacks, sometimes encouraged by state authorities (Daniels and Bassett 2002).

The mounting violence, which has spread throughout the Nakuru area, has made implementation of conservation impossible. Lack of security and uncertainty over future land control makes investment in landesque capital a gamble that few choose to take. Indeed, areas where conservation measures have been taken may become targets for violence, because such investments represent claims on the land. In any case, the conservation of Nakuru is a politicized problem and cessation of violence is prerequisite to any reforms in land management. This process, whereby ecological issues are politicized through local and regional conflict, and political questions and conflicts are increasingly cast in ecological terms, is another strong theme in political ecology.

The Argument

The environmental conflict thesis: increasing scarcities produced through resource enclosure or appropriation by state authorities, private firms, or social elites accelerate conflict between groups (gender, class, or ethnicity). Similarly, environmental problems become "politicized" when local groups (gender, class, or ethnicity) secure control of collective resources at the expense of others by leveraging management interventions by development authorities, state agents, or private firms. So too, existing and long-term conflicts within and between communities are "ecologized" by changes in conservation or resource development policy.

This argument is rooted in three fundamental lessons about social ecology, drawn from feminist theory, property research, and critical development studies. First, the argument works from an understanding that social systems are structured around divisions of labor and power that differentially distribute access and responsibility for natural goods and systems. Second, it reflects an understanding of property systems as complex bundles of rights that are politically partial and historically contingent. Third, it draws heavily on historical experience of development activities that shows them to be rooted in specific assumptions about the class, race, and gender of participants in the development process, often resulting in poorly formed policy and uneven results.

Social structure as differential environmental access and responsibility

Divisions of labor and access to productive resources mark human societies around the world. While the specific distribution of those divisions differs from society to society and across historical periods, the persistence of this pattern is universal. Different people, it has long been observed, are expected to carry out different kinds of work, and are allowed control over different environmental goods (Weber 1978, see especially vol. 1, chapter 2). So too, given the wide range of configurations, no single distribution of labor, access, and responsibility is natural or inevitable. Finally, from a normative political point of view, many such configurations are ethically and morally undesirable. Lack of access and opportunity and unfair distributions of labor burden are commonly rejected and so become the source of political struggles, as the history of women's rights, unionization, and civil rights in the United States has shown.

For political ecologists, the concern with this social fact is twofold. First, methodologically, an examination of environmentally oriented development activities or conservation efforts necessarily includes a careful census of (a) who controls what, (b) who is allowed to decide about what, and (c) who is expected to do what tasks. If conservation efforts alter ecological systems such that some productive resources are enhanced but others are hindered, the results will inevitably impinge differently on different groups, potentially creating or increasing conflicts and struggles. So too, any political or development effort geared at a specific group will hold implications for the environmental systems and flows governed by that group, gender, or community. Similarly, political efforts by different kinds of producers or managers, often at significant geographic distances, may be tied together by the mutual ecological systems they influence. Explicating, recording, and tracing these rights and responsibilities of the rich and poor, old and young, men and women, therefore, occupies considerable time and effort in political ecological research.

Second, this attention to division of access and responsibility constitutes one of the central normative concerns of most political ecological researchers. While divisions of labor are not, in and of themselves, in any way problematic, ecological change that unduly burdens some while benefiting others raises questions about alternative ways of doing things, and challenges the hidden costs of environmental conservation, remediation, or degradation.

Property institutions as politically partial constructions

More specifically, property institutions have proved to be complex bundles of rights, whose specific distribution has influence over trajectories and types of social/ecological change. The idea of property as a bundle of rights rather than a singular right is perhaps alien to many of us, especially those native to the United States, who tend to think of ownership as binary – either something is yours "fee simple" or it isn't.

In reality, no such system exists, even in apparently simple property systems like those of the global north. The ownership of things tends to be divided into a wide array of actual rights that may be exclusive to an individual or shared with a group. These include the separable rights to possess, use, manage, control income from use, and control capital value (Honoré 1961).

An individual may possess the right to manage land, for example, but not the right to income from its use. Such is the case of public land managers around the world. A community may have the right to use land, but the right to the income from its use may in part rest with another owner, as in the case of sharecropping systems universally. Someone may have the right to possess and use land, but not to transmit it through inheritance to his or her children, as where some environmentally oriented development easements seek to return land from use without expelling the current owner.

The complexity of property rights over natural goods and systems, especially in traditional societies, is an essential part of understanding social and environmental change, and its implications for land degradation, sustainability, and equity. This is especially true as many development efforts increasingly focus on creating more "rational" systems of property rights by making them more exclusive. Where a complex traditional arrangement, for example, might give local women the collective right to harvest from trees on land that belongs exclusively to a particular man, who in turn must open the land to local herders for dry season grazing, the institution of exclusive rights mean that this complex bundle would be collapsed into a single right, under the control of a single individual.

Such complex rights have evolved over long periods to manage the many temporal and spatial variations in the landscape (trees, crops, grasses) and the varied systems and divisions of labor in the local community. Privatization of rights commonly leads to resource conflicts, production losses, and increasing inequality.

As Meinzen-Dick et al. point out: "The widespread trend to privatize resources and to confer formal ownership to land, water, or trees, which has been promoted as improving economic efficiency and reducing transaction costs, too often cuts off more marginal users, and has particularly restricted women's rights to resources. More flexible tenure arrangements are more likely to accommodate the needs of multiple users of resources" (Meinzen-Dick et al. 1997, p. 1300).

Property rights over nature politically and ecologically mediate between differing users and ideologies of use. By examining their variations and their change, therefore, political ecologists get a better grasp on the relationship between environmental and social conflict (Emel et al. 1992; Emel and Roberts 1995).

Environmental development and classed, gendered, raced imaginaries

Finally, the environmental conflict thesis is rooted in a reading of the history of conservation and development informed by postcolonial and feminist criticisms (Chapter 3). Specifically, postcolonial analysis of history has demonstrated the way that development and environmental management initiatives, no matter how well intended, tend to be based on assumptions that are classed, gendered, and raced.

In particular, development plans tend to imagine the subjects of development – the local farmer, herder, or fisher – with assumptions about their outlook, behavior, and interests that reflect the socially situated imaginaries of the planner. This tendency has the potential to cause environmental conflict. Since these assumptions tend to view resource users and environmental decision-makers as monolithic, with shared interests, they tend to overlook the fact that the interests of different household and community members may diverge. Many important individuals and groups, because of their political marginality, are sometimes invisible to decision-makers, planners, and donors. Ignoring them does not make them vanish, however, and their role in managing, maintaining, or harvesting environmental systems usually becomes all too clear as conflicts erupt. Similarly, essentialist views of women, men, peasants, and herders create often bizarre expectations on the part of development planners, leading to differential investment and support for particular groups. There is, for all these reasons, a tendency for regional or global environmental management efforts to become enmeshed in local struggles, especially when outside authorities change the conditions in which people make a living.

The Evidence

There is a great range of case studies documenting these kinds of struggles. Some highlight the way pre-existing political differences become "ecological" – in the sense that longstanding struggles over social and economic power (e.g., labor movements, ethnic territorial disputes, or women's struggles for decision-making autonomy) are newly expressed or reframed as fights over the environmental (e.g., disagreements over conservation policy, finger-pointing over land degradation, or seizure of environmental goods). Conversely, other studies chart the emergence of new political divisions growing from existing ecological conflicts.

Agricultural development in Gambia

Despite regional variation, in Africa women constitute a large proportion of all farmers. West Africa in particular, one of the prehistoric breadbaskets of the world and a source of many of its important cultivars (Carney 2001), is a region where women cultivators have relative autonomy and have in many places historically held rights over property and the fruits of their labor.

It is also a region that has been consistently targeted by international development agencies for improvement of food and cash crop production. Experiments with cotton, peanuts, rice, and myriad other crops have turned many West African nations into grand experimental stations, often with disturbing results. Common to these efforts is an interest in facilitating agricultural intensification: an increase in the output of crop per unit of land.

Despite the importance of labor-saving technology in such intensification, it is often the case, as Carney and Watts put it, that:

> agricultural intensification is about getting people to work harder, a process that is social and gendered (getting *some* people to work harder than others) and that is typically coercive and conflictual. The manner in which labor intensification is negotiated and struggled over – that is to say, how agricultural intensification is played out through determinate rules of access to and control over resources – fundamentally shapes the character and the trajectory of agrarian change itself. (Carney and Watts 1991, pp. 652–3)

In her research into agrarian change in the Gambia, moreover, Judith Carney puts this conceptual claim to the test, evaluating how rice intensification schemes became the political and ecological point of change for the largely agrarian population, creating conflict and becoming the fulcrum of already existing struggle.

Gambia and the gendered land/labor nexus Gambia is similar to many African societies – including both matrilineal and patrilineal societies – where access and use rights of land and other productive resources are divided by gender. In the case of Gambia, a thin country following the course of the Gambia river, this gender division closely follows ecological zones created by the grade and flooding of the watercourse. Women have traditionally managed rice production (*Oryza glabberima*, distinct from Asian rice varieties), carried out in the periodically flooded plain, and control the harvest of their labor. Dryer upland fields were traditionally cropped with cereals like millet and sorghum, and managed by men, though traditional labor management meant some gendered mixing of labor and shared crop work. With the growth of markets for groundnut ("peanut"), dryland/upland production became more important in the colonial period of the 1800s. As a result, this separation of labor became more distinct, with women becoming primarily responsible for lowland subsistence rice and men for cash cropping peanuts.

More significantly, labor was divided both in terms of where labor was obligated to be performed, as well as who controlled agricultural products from different kinds of work. For the Mandinka people of the region, seasonal obligations to produce crops for the household (called *maruo*) are coupled with seasonal rights to produce goods for sale and retain the personal profits (*kamanyango*). Men and women each possess these rights and obligations in different places and at different times, although, under changing development initiatives, these became the source of increasing manipulation and conflict.

Specifically, the tendency for the colonial cash-crop economy to drive the food system towards rice imports (in what had earlier been a rice-exporting region, see Carney 2001) inspired development responses by authorities in the late colonial and independence era. Beginning with rice mechanization efforts by the British Colonial Development Corporation and later in the form of large irrigation projects, development efforts pressed for systems of rice export to undo the imbalances created under colonial-era cropping innovations. International development donors later joined these efforts with pressure for horticultural food crop development.

But the mechanization and irrigation of rice, and later the pressure for horticultural development to create food security (assumed for obvious reasons to be an apolitical and normatively desirable outcome) had serious and divisive implications. In particular, these efforts were increasingly used to place *maruo* labor demands on women's land and time, extending them to new areas and over longer parts of the year. Women were increasingly expected to produce household food through increased labor on land that had previously been used for their own *kamanyango* cash production. These demands came both from development authorities, who saw women as more cooperative, and from Mandinka men, who benefited from the increased harvests with no cost in their own time and labor (Carney 1993).

All of these changes precipitate conflict. The loss of *kamanyango* land and labor benefits means a loss of autonomy for women as well as a loss of cash, further

Box 9.1 Development undoing itself in Carney's "Converting the Wetlands"

"Most rural Sahelian people are forced to labor in brutal heat and humidity and survive on less than a dollar a day," Judith Carney explained to me in 2002. "I wrote 'Converting the Wetlands' because I wanted to show the significance of gender for understanding contemporary forms of agrarian transformation in the West African Sahel." These motivations, coupled with years of fieldwork, inform Carney's article (Carney 1993), which directs attention to development efforts that exploit long and grueling hours of women's agricultural labor, on land traditionally controlled by women, and which break down precisely because of a failure to understand and incorporate gendered rights to labor and land.

The most profound irony of the piece is that the development efforts Carney describes are explicitly intended to *help* women. These projects promised income benefits to women through commercialization of rice landscapes, which had already been planted with rice by women for generations. As Carney explains, these rice irrigation schemes:

> aimed to access skilled female labor for household rice production while denying them the benefits from their labor . . . An acute conflict developed between the productivity objectives of the project managers/donors, the labor demands placed on household producers, and potential profits from rice growing by participating households [and] caused the project to be regarded as yet another instance of the presumed inability of African producers to manage complex irrigation schemes when in fact, the problem was the project designers' willful disregard of the social organization of production, rural poverty and regional resource struggles over wetlands.

In this sense, political ecology is not necessarily an anti-development research project. In many cases, research seeks to explain how development undoes itself, through habits of thought and indifference to local reality.

"The objective is not to take money from men to give to women. My hope is to find ways to improve food security and economic options by exploring patterns of land use and development that build on autochthonous traditions and farming systems."

Imagining that things might be better, as Carney does, is therefore prerequisite to the most important research.

creating a breakdown in household cooperation and reciprocity. According to Carney, women increasingly demand cash compensation for their losses, withdraw their labor, and politically challenge agricultural schemes (Carney and Watts 1991).

In sum, changes in cropping and technology mean changes in property, which mean changes in labor and labor burden. Some community benefits are someone else's personal costs. In this case, as in many, the price of development was paid by women, specifically as a result of the social ecologies described above, where labor and rights are socially stratified, property rights are fluid and complex, and development authorities make gendered assumptions about labor and land which later strike political sparks.

Such development efforts persist nonetheless, despite setbacks and failures. Nowhere is this clearer than in the "women in development" (WID) movements of the 1980s, which swept across sub-Saharan Africa as a new paradigm that would take women seriously. Serious attention to women's economies does not erase the underlying gendered struggles for power in households and communities, despite the best intentions.

Efforts by Save the Children and other NGOs and mission groups in Gambia during the 1980s, for example, attempted to bring funding and technical attention to women's horticulture, following a spirit of WID. Progress was followed by a dramatic backlash, however. Subsequent non-governmental interventions concerned with recruiting women for agroforestry were seized upon and manipulated by local men and directed to land that had often previously been garden plots. As men predominantly control tree resources, this effectively allowed them to simultaneously seize control of garden lands and of women's labor. In this case, two separate strands of progressive environment-based livelihood schemes – gardening and agroforestry – were implicated in an ongoing gender struggle where men and women used each to economically bludgeon one another (Schroeder 1999).

An immediate lesson is that an increased focus on *women* in development by no means represents a better understanding or engagement with *gender* and the power relations inherent in a system with a strict division of labor. The case therefore reveals the patriarchal assumptions of environmental management and development efforts. Indeed, the idea that "saving the children" is best accomplished by supporting women's efforts exclusively appears highly problematic in retrospect. At a more abstract level, the case further documents the serious local politics of production into which all development interventions must inevitably become embedded – making decisions about gardens or trees in development is to make a decision about local power. Much the same can be said of public lands management in the western regions of the United States.

Land conflict in the US West

In 1993 the governor of the state of Colorado established the Sangre de Cristo Land Grant Commission in an effort to settle a dispute of 38 years' standing over a 77,000 acre tract of land known as the Taylor Ranch. Following the new model of democratic, collaborative, and cooperative dispute settlement now favored in many parts of the US West, the commission sought to resolve ongoing conflicts between local Hispano residents, who settled in the region in the 1800s and who claim usufruct rights to the land, and the private owner of the land. In 1997, however, and after several years of wrangling, the commission's effort proved a failure; to the consternation of the community and many observers, collaboration failed to reach a community-wide agreeable solution to the conflict.

As Randall Wilson observes in his analysis of the San Luis Valley conflict, however, it is the very nature of the Sangre De Cristo Land Grant Commission and the character of its intervention that politicizes the region's ecological relations. The commission operates from a conceptual arrangement that erases the local Hispano community and its quest for use rights, which it was ironically designed to champion. Specifically, the commission, in its mission statement, its goals, and its strategies, justified change of control over the Taylor Ranch land in terms of its value as a "natural" place, and so one best managed under the scientific knowledge and authority of the state, as well as in terms of its "traditional" cultural value to the Hispano community. This ran counter to Hispano community desires for productive access to the forest for hunting, wood collecting, and grazing.

By sidestepping the Hispano community's desires for a nontraditional and productive landscape, and championing a "natural" and culturally "authentic" use system, and by making state managers central to the management of the land, rather than local hunters, wood collectors, or grazers, the commission conceptually dismissed the marginal community it had intended to aid. By centering on state managers (rather than Hispano land users), the commission gave fodder to opponents who claimed the decision reflected an erosion of private property rights. It also meant that the commission's proposal could be undermined by opponents' claims of *political bias*. By making the case for state intervention in the "objective" terms of scientific state-based resource management, the mere accusation that the commission might be motivated by a political effort to right historical wrongs, cast the entire enterprise in doubt. The commission's insistence, moreover, on focusing on conservation and state expertise (as in Chapter 8) only exacerbated local discontent, fomented not in the protection of nature, but of use rights. Had the commission set the terms of its intervention on equity, use rights, tenure, or access, the outcome might have been quite different.

The conflict is typical of many ongoing struggles in the US West, where the long-time residents of areas in and around public lands increasingly struggle to defend historical access and use rights in these lands, finding themselves at odds with environmental groups, new residents, and state agencies (McCarthy 2002; Walker 2003). It is also typical of global ecological conflict scenarios (like those in West Africa), where decentralized and facilitated land management and dispute settlement are increasingly implemented to reach peaceful outcomes, but often with undesired consequences.

But it more generally calls attention to shortcomings in environmental interventions by development authorities, including environmentalists, around the world. The Sangre de Cristo Land Grant Commission operated from a classed and raced conceptual system. The commission's symbolic system was incongruent with the character of the pre-existing environmental conflict. The practical result was one desired neither by the state (who expended costs and political capital to no effect), by environmentalists (who left the land in the hands of the owner, an aggressive logger), or by local Hispanos (whose 100-year grievance received no fair hearing). The more general implications of the case, as for well-intentioned efforts like those of Save the Children in the Gambia, is that by failing to acknowledge the political character of management, with its equity and power implications, the stage is set for failure.

Evaluating the Thesis

These kinds of environmental conflict cases probably represent the greatest bulk of political ecological research, and this environmental conflict research, among the wide range of approaches in political ecology, has made the greatest practical impact. Showing the distributive justice outcomes of environmental and economic change, this research makes it increasingly difficult for planners, states, or lending agencies to ignore the churning regional struggles into which any environmental management scheme will inevitably entangle itself. This detailed research, however, does more than simply assert the somewhat banal overall conclusion that ecology is political. It further demonstrates and reveals:

- that the fundamental hinge-points between human beings and the environment are not restricted to technologies of production (tractors, factories, automobiles) and levels of consumption (population and affluence), as is commonly asserted in apolitical ecology, but extend to distribution, access rights, and the division of labor
- that despite the very material character of environmental struggles around the world, it is often concepts and constructions of community and nature that propels or suppresses conflict
- that the equity and sustainability of environmental management is not dependent on the scale of environmental governance (local, federal, community), as is sometimes asserted in anarchic or romantic localism, but depends rather on the specific arrangements of differing groups in an ecological network
- that increased attention to poverty and women in environmental issues, a hallmark of new development initiatives, is not the same as attention to class or patriarchy

More generally, the work presents a serious challenge to single-objective environment and development initiatives and encourages a view of the landscape through many lenses, as claims on ecological system components overlap in a complex environment. This also suggests an ecological critique of exclusive property rights, at least as they are popularly understood in the global north.

Stock characters and standard scripts

There is a tendency in this work, however, to treat many groups or "stakeholders" (a problematic term in itself, originating in business management theory) as categorically real. The work tells us about "men"/"women" or "herders"/"farmers" for example, in ways that do not seem to ring true, especially given the fluidity of the resources and rights so carefully documented in the research. These categories and groups themselves develop out of fluid ecological positions as much as the other way around. Political identities are not simply a product of social/cultural values, but also of rice fields, pasturage, and ranch boundaries. As these systems change, as a result of environmental, political, and economic shifts, we should expect the categories themselves to shift, and for these groups to dissolve and re-form before our eyes. This does not always seem to be the case in political ecological research, however, where stock characters often walk on and off stage as if on cue.

In much the same way, these investigations of the behaviors of individuals or groups tend to provide a rich analysis of the contextual forces that put groups at odds with one another and create new distributions of rights, but with a far less clear account of how people actually respond, change their behaviors, or alter the landscape. The wide range of adaptations available to people and the dynamics of their responses receives too little attention in many otherwise excellent analyses. The result is that the environmental conflict literature sometimes reads like a fairly standard script. It also means leaving aside many important questions: What accounts for the specific adaptation choices made under conditions of conflict? Along what lines are new environmental decisions and behaviors formed? What are the material environmental outcomes of conflict? To ask these questions and address such responses in a more than descriptive way might of course require new tools; decision analysis is scarce in political ecology.

These two issues further undermine an important potential project: explaining when conflict *doesn't* happen. Indeed, the inevitability of environmental conflict is by no means assured. For example, as some political ecologists have recently pointed out, tenurial systems where men's and women's responsibilities and access are carefully split can be potentially complementary and negotiable, allowing conflict to be averted in many cases and sustainable use to be achieved (Rocheleau and Edmunds 1997). But if decentralized and collaborative schemes do not in and of themselves assure sustainable outcomes, what does? To address this question means that we must better understand (1) how groups are not just situated but actually *formed* in ecological networks, and (2) how people *respond and adapt* with new social and ecological categories and strategies, influenced by knowledge, context, and political process.

Research Example: Gendered Landscapes and Resource Bottlenecks in the Thar

Whether or not conflict is inevitable, it is common, especially where large-scale environmental interventions lead to differentially distributed impacts. This is nowhere clearer than in the zones of green revolutionary agricultural change in India. The innovations of the green revolution – including high-yielding varieties of important cultivars, electrified groundwater pumping, and increased use of chemical inputs – have unquestionably increased the amount of food produced. The way in which these benefits have been distributed and the degree to which they have entailed hidden costs to some groups is as yet quite unclear, however, as is the degree of conflict that may arise from disparities in costs and benefits of agro-ecological change, especially between men and women.

Several factors affecting women's relationship to resources in India should be noted. First, despite differences based on class and caste, women are often involved in all spheres of the household. Agricultural labor is often heavily female; in the Jharkand region of Bihar, for example, for each 80 days of human labor required per acre annually, female workers supply 65 (Sharon and Dayal 1993). Women are often simultaneously responsible for procurement of fuelwood and other minor forest products, acquisition of fodder, herding of animals, child care, cooking, cleaning, and home construction and maintenance. Also, despite women's labor outside the home, men control much of the decision-making in agriculture. Because most communities follow a patrilineal and patrilocal pattern of social organization, young women typically marry into household economies where they become the least powerful and least well-established participants (Liddle and Joshi 1986).

Aggregate analysis of the impact of the green revolution on women points to a few trends. Despite increases in food supplies resulting from intensification, women experience disproportionately poor access to nutrients and food (Das Gupta 1987, 1995; Messer 1997; Cassidy 1980) and the gap between male and female agricultural wages (of the order of 30 percent) is increasing (Singh 1996).

But a more fundamental question suggests itself: Do the specific ecological changes in green revolutionary production systems change the labor burden, resource rights, and environmental access of women? Here, secondary data on wages and calories become of limited value, and research must turn instead to the daily ecologies of the household and the specific dynamics of land use and land cover change.

My research in the Thar region of Rajasthan (introduced in Chapter 7) was extended to try to understand the changes that have occurred in women's daily workloads and in their access to and control over resources as a result of intensification. Rajasthan is instructive, since the adoption of modern intensive systems is widespread and gaining momentum. Whereas only 2.6 million hectares of land in Rajasthan were under irrigation in 1976, 5.5 million hectares were irrigated in 1999. The area under high-yielding varieties (HYVs) has also climbed; 255,000 hectares of land were under improved varieties of bajra (pearl millet) in 1976, while nearly 1.4 million hectares of HYV millet were grown in 1999. HYV wheat production climbed by more than 250 percent between 1976 and 1999 to a total of 2 million hectares. More food is being produced, to be sure, but does this lead to conflict over resources and differential burdens in the community?

Methodologically, answering this question required:

- determining differential land uses and rights and discovering what species different people depend upon
- tracking changes in availability and determining how the availability of these resources has changed over the period of the green revolution
- evaluating divergent impacts and weighing the benefits against the costs of environmental change

Determining differential land uses and rights

As explained previously, the challenge in determining people's land uses and rights is made difficult by the "ordinariness" of daily actions; it can be difficult to query people about things they do every day, since those things seem trivial or taken for granted. This is coupled with the fact that use rights are often unstated, unwritten, and largely "understood" in a way that often makes them hard to articulate. For that reason, observation is usually preferable to interview, and certainly to questionnaires.

This being the case, methodologically, I proceeded by following people on their daily routines, asking occasional questions, and participating when I could – carrying sheaves of dried hay, for example, a task that is exhausting and somewhat itchy. In particular, I asked people about the important non-domesticated species that they used and about the places where they obtained them. This informal and intensive work was supplemented by a survey of men and women, with each individual listing the species important to production.

The results point to species and land use areas that are differentially important for women and men. Specifically, forest and fallow lands provide important species inputs into household and livestock production, and the collection of these represents a significant amount of time and effort on the part of women: the reproductive labor burden of the household. These surveys also revealed precisely when certain lands were important. In particular, traditional forest lands (*orans*) and the more marginal stony and sandy lands around the village were most important in the growing season (when agricultural land is under crop), while fallow land is most important in the dry season, when productivity of the more marginal lands is exhausted.

There are several problems with this analysis. First, the class and caste balance of the sample is extremely biased. This is a result of very different social norms for low and high caste women in the Thar region as well as some of the same caste-based issues that created boundaries in previous work (Chapter 7). Traditionally, lower and middle caste women (*meghwal*) as well as women from pastoral groups (*raika*) are easier to approach, spend more time out of the house, and are involved in wage labor that involves travel and interaction outside the village. Also, my primary co-investigator was a member of the *meghwal* community, and our best contacts came from his associations. As a result, women from these groups became primary respondents and some groups, including and especially those from elite *rajput* families, largely went unsurveyed.

Tracking changes in availability

Once the timing and spacing of women's and men's resource use was roughly determined, it was necessary to determine changes in the coverage and availability of those lands, at least since land reform occurred in 1955. Since forest and fallow lands are especially important for women, the change in coverage of these lands can be queried in only a general way from Indian land use records, which do show the general decline of "waste" lands in the region, a category including all land that could be used for cropping but currently is not. This land has slowly fallen under the plow as double-cropping, irrigation, and chemical inputs enabled expansion of planting in time and space. This confirms the general notion that important lands for reproduction of the household have been lost, even as productive resources have expanded.

These general trends are relatively meaningless, however, especially since they provide no indication of the seasonal availability of resources and because the categories are aggregated ("waste") to include many different land uses and land covers, whose specificity are important to making a living. To get a more detailed and disaggregated sense of land use/cover change it is necessary to turn towards local records.

These records, *misl bandobast* records from land settlement in 1955 and more recent *jamabandi* records, exist for every village in Rajasthan. These are paper documents, threaded and written by hand, and keyed to large cloth maps kept by a village record keeper – the *patwari*. While the contemporary *jamabandi* records are kept in English and Hindi, the *misl banadobast* are typically recorded in Urdu, the administrative language of the Mughals still in use during the time of the British Raj and even in early independent Rajasthan. More problematic, however, is the fact that most of these records are stored in derelict buildings, in semi-organized piles, bound in bundles, coded by color and often-obscure markings. Though the arid conditions of the

Figure 9.1 An important resource for political ecologists: detailed, village-scale, historical land cover data, using local categories, exist for many areas, but are in an advanced state of decay

region help to preserve the paper somewhat, decay has set in for many of these records, and some are in their last generation of usefulness (Figure 9.1). This is a far cry from the condition of some colonial records (described in Chapter 8), and presents some important challenges for research.

Even so, it is possible to assemble a record of change for a sample of these villages. By coupling the more detailed land cover data with measurements on the ground, interviews, and survey work, a rough profile of resource availability becomes clear. The resource calendars shown in Figure 9.2 are based on the mean land coverages described for 1955 and 1994 in village records for a sample of 29 randomly sampled villages (see Robbins 1998a). The vertical axis represents the spatial average coverage of each land type per village in hectares, while the shading denotes the quality of resource availability based on species coverage across the year.

Evaluating divergent impacts

The 1955 arrangement shows a system in which the loss of resource land to cultivation during the growing season (July to October) is offset by the availability of

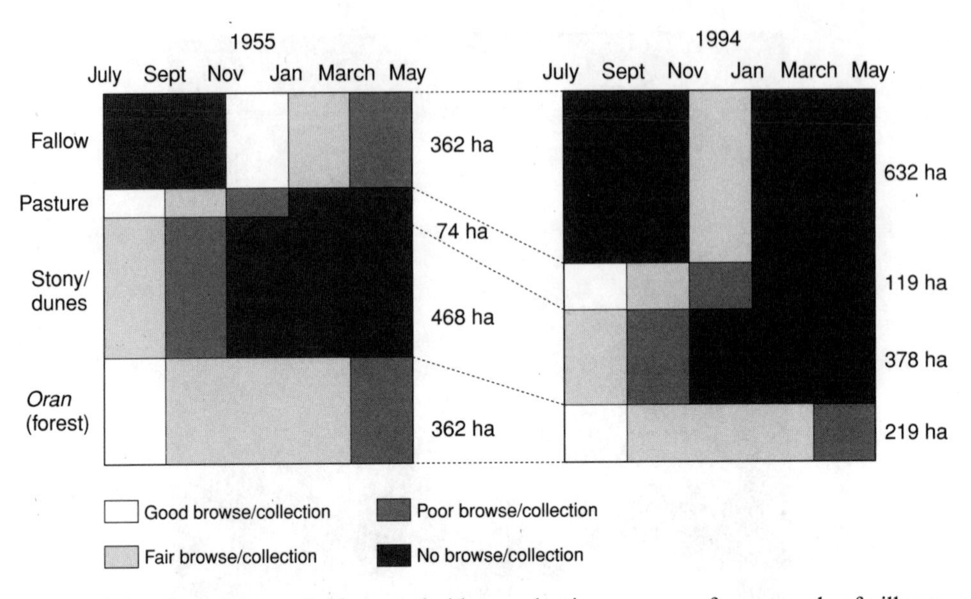

Figure 9.2 Change in available household reproductive resources for a sample of villages in Rajasthan between 1955 and 1994 (Reproduced from Robbins, figure 25.4)

alternative resources, especially forests (*orans*) and other "marginal" and "waste" lands: stony lands, pasture, and sand dunes under grass and scrubby vegetation. When these lands become less productive in the dry season, fallow lands become available for use and harvesting. Women's labor, therefore, solves the *temporal* problem of scarcity by shifting *spatially* to obtain key reproductive resources.

By 1994, however, forest lands and dunes had been lost to enclosure and cropping. The *oran* lands in the region, though a key resource, had been enclosed for cropping. Moreover, as land is increasingly cropped twice and three times annually, dry season fodder and fuel resources become scarcer as well.

These changes mean an increasing labor burden for women. Daily fuelwood and fodder gathering activities are reported as more time-consuming and more difficult for women in the survey villages. Girl children are increasingly involved in these activities in many households. Put simply, an expansion of cropping and an increasing flow of resources into the productive (and largely male) sphere of the household has meant a decrease in women's resources and an increase in women's work. Development is by no means a win-win prospect for locals, therefore, and portends conflict.

Have these changes meant the removal of girl children from school? Have they had an impact women's health or autonomy? Has this trend resulted in changing power relations within households? Has it led to an increase in conflict? Have women negotiated new labor relations as a result of changing resources? Have their species priorities or management of the ecosystem changed? What are women and men doing now? I was unable to answer these questions with the limited time and resources available, but they remain an urgent direction for future research.

10

Environmental Identity and Social Movement

Skull Valley, Utah, is the desert home to the Goshute Indians, and has historically played host to a range of toxic dumps. The 18,000 acre reservation 45 miles southwest of Salt Lake City is proximate to a military testing range, a hazardous waste incinerator, a hazardous waste storage facility, and a refinery that is the source of a majority of the nation's chlorine gas pollution emissions. In short, Skull Valley is a dump. But more than this, it is a sacrifice zone where, for more than a century, a dwindling Goshute community, who even by native North American standards are a marginal group, has paid the costs of military industrial development unwillingly.

The passage of the Nuclear Waste Policy Act in 1982 and its amendment in 1987 fundamentally, if indirectly, changed the stakes in this region, however. This act, which slated the construction of a repository site for high-level nuclear waste at Yucca Mountain in the Nevada desert, meant that the long-discussed plan of transporting radioactive waste for long-term storage in the US West was at last at hand. Such a plan further necessitated the construction of a number of smaller, temporary, monitored retrievable storage (MRS) facilities. When the Department of Energy failed to close any deals with communities who would agree to host such a site, private utilities opened their own searches and negotiations and in 1996 the Goshute came forward to arrange a leasing contract (Ishiyama 2003).

As explored in the research of Noriko Ishiyama, the Goshute staked out an unusual and dramatic environmental justice and livelihood argument and performed an end-run around Utah's state authorities, who were dismayed, and around environmentalists and other concerned observers, who implored the Goshute to reconsider. In their efforts to arrange the contract, Goshute elders pointed to the desecration of the valley throughout history, emphasizing that the degradation and marginalization of the region went unremunerated for a century. Arranging the creation of an MRS site on their land was thus an assertion of traditional sovereignty over the land. It was also an attack on outside groups that would on the one hand insist on the traditional and preservationist romances of native communities ("green" Indians) while doing little to attend to the crisis of underdevelopment and marginalization in the region (Ishiyama 2003).

As Ishiyama points out, the historic conditions of marginalization and degradation provide the necessary context within which to judge whether a decision to locate a hazard is just or unjust. In the absence of such a history, the locational decision for high-level waste must appear unjust, with the tribal government complicit in targeting marginal communities for high-risk conditions unfairly. Set against the backdrop of a continually trashed and exploited landscape, and in a context where other communities attest to speak for the Goshute, a more complex picture emerges, where the siting effort appears in a postcolonial context of struggles for identity, sovereignty, and power over the landscape. As a result, similar native efforts to secure MRS sites have been strikingly common (Erickson et al. 1995).

Native movements that challenge the collective wisdom of sympathetic observers therefore establish struggle on a double front. By asserting sovereignty and control, native people set the terms of development for investors and firms who seek to develop native lands. But by selecting developments that seem non-native (waste dumps as only the most prominent example), native peoples reject the romance and sympathy of environmentalists and other groups who would prefer to see different and more "traditional" land use practices.

Exploring these demands for autonomy around the world, such as those of the Zapatistas in Chiapas, Mexico, or the Miskito of Honduras and Nicaragua, and analyzing the livelihood conditions, moral economies, and political exigencies of regional life that propel them, forms the core of another fundamental body of work in political ecology: environmental identity and social movement research.

The Argument

The environmental identity and social movement thesis: changes in environmental management regimes and environmental conditions have created opportunities or imperatives for local groups to secure and represent themselves politically. Such movements often represent a new form of political action, since their ecological strands connect disparate groups, across class, ethnicity, and gender. In this way, local social/environmental conditions and interactions have delimited, modified, and blunted otherwise apparently powerful global political and economic forces.

In a sense, therefore, the environmental identity and social movement thesis is the reverse mirror image of the degradation and marginalization thesis (Chapter 7); where exploitation leads to the simultaneous destruction of productive resources and of local producers, it simultaneously draws together otherwise disparate communities and interests into collective awareness, and so to collective action. Here, communities assert their identity through the way they make a living. Drawing on a paradigm case of the struggle for rights of *adivasi* (indigenous tribal) people in Jharkhand, India, Parajuli (1998) has described this as a form of "ecological ethnicity." This argument draws upon three groups of tools in human ecology introduced earlier, including (1) the problem of unevenly distributed risks and environmental justice, (2) theories of peasant action, and (3) postcolonialism and ecology at the margins.

Differential risk and ecological injustice

A well-established record in environmental justice research has demonstrated that the risks and hazards associated with many development practices are disproportionately distributed to communities with less political and economic clout, especially communities of color. In the United States it is well established that hazardous waste is disproportionately likely to be deposited in minority communities (Bullard 1990), as an obvious example, and that pesticide exposure is disproportionately acute for migrant farm workers (Pulido 1996).

This branch of political ecology proceeds in recognition of that sordid and uneven track record of development and, like related research in environmental justice, further accounts for the way in which people and communities actively respond, adapt, and seize control of such conditions. Research into environmental identities and social movements explores the ties between specific environmental system components, the livelihoods they support, and the risks presented by particular development practices. This means that research often raises the possibility that changes in livelihoods coupled with *specific* controls over development (how dams are built, where value from mining goes, how industrial locations are decided) may yield very different and more progressive outcomes.

Moral economies and peasant resistance

This work also seeks to explain the conditions under which such movements form, unify, and mobilize. This question is far from trivial since many unjust conditions spark little or no political response and various systems of exploitation have sometimes proceeded quite smoothly with little or no resistance from subsistence producers.

The theoretical tools to address this question in part lie in the foundational work of peasant studies on moral economies. As noted previously (Chapter 4), prevailing conditions of subsistence give rise to systems of reciprocity to reduce risk and moral expectations about shared burdens. Such systems can survive and adapt to a range of livelihood shocks, from both economic and environmental sources. But when the structure of that livelihood system is challenged by fundamental changes in the way labor is remunerated or risk is distributed, social mobilization becomes more likely (Scott 1976).

Contemporary struggles go beyond rebellions against the overextraction of harvests and taxes (the central preoccupation of traditional peasant studies researchers). They turn instead on the way livelihoods are challenged and violated on a more general and regional scale by modern forms of development practice, like large-scale displacement, significant shifts in credit, and promulgation of new technologies. So too, such regional changes may unite communities that have traditionally been divided. The "ecology" in this emergent "ecological ethnicity" doesn't necessarily refer to natural systems, therefore, as much as to the linkages between various individuals and groups, and the livelihood threads that hold them together. These linkages are what make such extended communities so vulnerable to certain forms of development, as where a large dam-building project inundates the fields of both rich and poor. But such linkages also make communities potentially powerful, since they have the potential of acting in concert.

So contemporary peasant upheavals represent a form of social movement, which geographer Alan Scott defines as:

> a collective actor constituted by individuals who understand themselves to have common interests and, for at least some significant part of their social existence, a common identity. Social movements are distinguished from other collective actors, such as political parties or pressure groups, in that they have mass mobilization, or the threat of mobilization, as their prime source of social sanction, and hence, power. (Scott 1990, p. 6)

Postcolonialism and rewriting ecology from the margins

It is the key concept of individuals "who understand themselves" that underlines another key feature of the environmental identity and social movement thesis. This is the articulation of social action and environmental conditions on the terms of the disenfranchised or marginalized themselves. In this way, the thesis draws on postcolonial notions of the "subaltern" subject (Chapter 3).

Recalling that accounts of social action and social history, especially in underdeveloped places, have historically been written from a position of colonial privilege (foreign academics, state managers, etc.), a political ecology of livelihood movements represents an effort to make space for the forceful claims and actions of marginal communities. While these movements can be heard clearly enough on their own – consider the global press coverage of the Zapatista movement in the mid-1990s, which needed no "help" from academics – these efforts can help to make such claims a part of more general ecological inquiry. In other words, although *adivasis* in Jharkhand have independently built a solid case for local forest degradation and have managed to secure land rights through political action (Parajuli 1998), this does not mean that American forestry science will change to account for such knowledge. By building a political ecological record of such movements and claims, research does some work towards both validating local accounts and challenging dominant ways of seeing economic and ecological change.

The Evidence

Case study evidence on livelihood struggles is growing. Overlapping with more general research into social movements in development studies and rural sociology, the political ecology of livelihood struggles differs in that it attends to a greater degree to the material details that make up livelihoods (crop choices, labor rotations, seed storage, etc.) and the way those specific material conditions present limits and opportunities for groups as they organize, struggle, and seek to define themselves and their ways of life. Research in the Andes and Uttarkhand, India, is typical, in that it underlines this approach, even while showing some of the problems inherent in such research.

Andean livelihood movements

The highlands of Ecuador are a perfect laboratory for research in political ecology. The agricultural systems of the highlands (at 2,500–3,700 m) are remarkable in themselves, since they represent a set of serious constraints in terms of soil, slope, and climate. The degree to which the cultivation of potatoes, maize, and barley yield surpluses and stable populations, indeed supporting prehistoric empires, is a topic worthy of examination in itself. So too, the postcolonial history of the area presents interesting problems. The high-altitude regions are disproportionately occupied by indigenous *quichua*-speaking peoples and mixed indigenous–mestizo communities, historically linked to haciendas (large feudal landholdings) as laborers, but also holding their own small subsistence plots. With the breakup of the haciendas in the 1970s, land redistribution led to subdivision and accumulation. The resulting highly stratified state of landholdings in the region produces labor, class, and ethnicity dynamics of great complexity (Bebbington 1993; Jokisch 2002).

But this corner of the planet, with its peculiar history, is in many ways one of the most globalized and integrated regions in the world. The area has been a target for global development aid and technological diffusion since the dawn of the green revolution more than 30 years ago, and technicians, extension agents, and development officers have been crawling across the mountains for at least that long. So too, whole highland villages have been depopulated as migrants travel from this area to places like Queens, New York, and Barcelona, Spain, sending remittances with values in the millions of dollars annually. Thus, if there ever were a place to study the global–local linkages of survival, adaptation, and upheaval, this would be it (Jokisch 1997).

The region is also notable, however, for the way in which social movements and ethnic identity have been closely intertwined with changing crops, technology, and labor relations in the last few decades. Like indigenous movements in Ecuador more generally (Perreault 2001), these have demonstrated impressive variability, as a result of varying local ecological and political conditions. Even so, the rise of indigenous livelihood movements and the peculiar articulation of ethnic identity in the landscape point to some general patterns in the region: indigenous movements often embrace modernization, but on their own terms.

Modernization and identity The highlands is a region where livelihood movements do articulate local concerns, which grow from traditional agricultural practices, but do so in ways that utilize contemporary agrarian technologies to enhance survival and surplus. Anthony Bebbington, whose research within and about NGOs in the region has tracked the history of these movements, describes the way traditional highland *quichua* speakers (known as *runa*) have increasingly pressed for autonomy, rights, and land reform to break up haciendas and redistribute assets to the poorer communities. These indigenous movements, based in a notion of collective identity, not only successfully forced local land redistribution in the 1960s and 1970s, but also led to the creation of larger political federations, increasing the force and prominence of indigenous communities in national politics (Bebbington 1993).

The success of these movements, however, has ecological implications. Land has been subdivided and hacienda pasturage has given way to cropping. This means increased intensity of production on small plots coupled with a decrease in animal nutrients for the soil, leading to decreasing yields and erosion. Simultaneously, increasing integration with global commodity markets means new crops, increased migration, and tighter margins. Non-governmental and church organizations working in these regions have responded by offering packages of modern high-input agrochemical technologies to local indigenous producers. This green revolutionary approach was fostered in spite of an avowed interest on the part of these organizations to protect indigenous technologies and knowledges (Bebbington 1996).

But Indian federations, with 30 years of experience in articulating community needs, have responded to these technological opportunities in surprising ways. Responding to the realities of market integration and the disinterest on the part of local producers for entirely traditional approaches to production, they have begun to embrace the use of agrochemicals, especially fertilizers. By incorporating some, though not all, green revolutionary innovations to maintain crop yields, outmigration is reduced, leading to enhanced community cohesion. As Bebbington explains: “modernization, far from being a cause of cultural erosion, is explicitly seen as a means for cultural survival” (Bebbington 1996, p. 101).

Such efforts at modernization fly in the face of those accounts that point to the risks of the green revolution, both for dissolving traditional social structures and undermining the sustainability of ecosystems. Modern technologies often displace labor and create demands for cash that reconfigure moral economies, thereby, it is argued, tearing at ethnic identity, community cohesion, and local empowerment. So too, the high-input systems of modern agriculture have also been shown to be demanding on soil and water resources, leading to failure in the agrarian system over the long term.

But the Andean case suggests that such predictions are somewhat premature. Bebbington puts it simply: “while agrarian modernization led to the erosion of some ‘indigenous’ cultures, this need not be the case: it depends on how the rural poor are able to incorporate and use modernization” (Bebbington 1996, p. 90). In other words, livelihood movements need not be “traditional” to be effective, and by controlling the conditions under which introduced technologies and outside forces act, such movements allow traditional communities to thrive.

Box 10.1 *The Violence of the Green Revolution*: Shiva's big picture thinking

The Violence of the Green Revolution (Shiva 1991) is an extremely useful polemic. In it, Vandana Shiva, a physicist, philosopher, and feminist, explains the simultaneous eruption of environmental crises and social/cultural upheaval in the Indian Punjab by directing attention towards the high-input monocultures of the green revolution. Agrarian technological and development changes since the late 1960s, Shiva argues, have led to water conflicts and ethnic strife, even resulting in large-scale mobilization of nationalist separatism amongst the ethnic Sikh majority of the Punjab.

As Shiva points out (though doesn't exactly prove), green revolution wheat production in the Punjab has contributed to land consolidation, conflicts over scarce water, and unsustainable inputs of chemical fertilizer and pesticides. These technological innovations (high yielding varieties of wheat, tractors, and herbicide/insecticide packages) and their concomitant institutional services (extension and research, formal systems of capital lending, centralized water management) have indeed turned agrarian production in north India upside-down. Even those technical experts who have long supported the green revolution have begun to admit that the system has tapped-out biodiversity, soil productivity, water, and human capital in India, leaving them scratching their heads for new methods to sustain high yields (Lal et al. 2002).

The broad brush with which Shiva paints, however, results in a number of overgeneralizations and mistakes. It is probable that Sikh nationalism was around before the green revolution era, and conflict over water in the region is not novel, since it dates back at least to the time of partition. The book's assault on "reductive science," moreover, and the development system behind it, appears to speak on behalf of the agrarian poor of the Punjab, but detailed political ecological fieldwork has shown the creative, if cautious, incorporation of green revolution technologies by smallholders throughout the world. So too, Shiva's distrust of development, defined in the most general terms, extends both to private chemical and seed firms and the Indian state, entities with very different motives and progressive possibilities.

The risks and benefits of polemic in political ecology must therefore be weighed. To grab attention and focus it on legitimate crises is an unequivocal skill, which too few researchers possess. The green revolution has strained ecological systems and resulted in outrageous and violent rural social stratification, and any reader of Shiva's book is forced to consider these issues seriously. The book is also notable in that it is easy to read, honest in its political goals, and directed at an audience who may not be familiar with the green revolution, its history, or its many implications. Attempting to leverage so large an argument and attributing so much causal power to forces defined so generally, however, represents hurried scholarship, which detracts from political ecology's rich research tradition. The lesson of *Violence of the Green Revolution* is, therefore, to wield political ecology's heavy political hatchet with care, and see to it that the analytical blade remains sharp.

Evidence for this effect can be seen in other Andean contexts. In highland Peru local peasant communities have innovated a range of cropping practices to take advantage of regional and global markets, even while resisting state authorities and non-indigenous groups (Zimmerer 1991). The extension of labor migration from Ecuador, and the massive remittances that such globalized livelihoods allow, similarly has not led to a radical modernization of cropping systems. Rather, remittances are poured into the acquisition of land and the construction of new homes, since conspicuous consumption of land is important in the regional cultural economy (Jokisch 2002). Thus modernization, mediated by social livelihood movements, has meant new forms of "ecological ethnicity" that are by no means strictly traditional.

Hijacking chipko*: trees, gender, livelihood, and essentialism in India*

Perhaps one of the most emblematic, and therefore misunderstood, of all livelihood movements around the world is the *chipko* movement of the Indian Himalayas. This movement has become so enshrined in global environmental consciousness that it is commonly mentioned and described as a model by activists and observers who have never even set foot in India. The lesson that might be drawn from *chipko*, however, extends beyond understanding what a livelihood movement is and how it works. Rather, the case holds implications for explaining how actions become emblems, myths, and ideas, sometimes allowing us to ignore and forget the communities that launched such movements in the first place (Rangan 2000).

The story of the movement, as it is commonly told – I have seen it in places ranging from children's television programs in India to Sierra Club publications in the US – is that local people living harmoniously in the Garhwal Himalayas of India witnessed a state-sponsored process of reckless commercial deforestation in the late twentieth century and, rather than stand for it, in 1973, began wrapping themselves around the trees of the region, defying the axe-men charged with felling the forest. As the story goes, *chipko* was so successful that it came to serve as a rallying point for environmental movements around the world.

Women's movement or peasant movement? The deluge of political ecology that grew from this movement represents a kind of cottage industry for critical scholarship. Questions that emerged from these events focused on what the movement can tell us about emergent environmental protest, about the relationship of women and peasants to nature, and about the possibilities for resisting modernization, globalization, and capitalization of nature. Of the mass of scholarship, most notable are the works of Vandana Shiva and Ramachandra Guha, both veteran scholar activists, and both dedicated to imagining alternatives to unsustainable ecological practice. But in their quest to characterize and celebrate *chipko*, each diverged in their interpretation of events.

For Vandana Shiva, perhaps the most dedicated and articulate spokesperson for global ecofeminism, the *chipko* movement was fundamentally about colonialism, women, and gendered relations to nature. As recorded in her ecofeminist classic *Staying Alive* (Shiva 1988), the movement was an expression of a resurgent "feminine principle," an essential relationship between women and nature. After the colonial destruction

Box 10.2 A road less taken in Zimmerer's "Wetland Production and Smallholder Persistence"

Political ecologists have a long and abiding interest in resistance: people's conscious opposition to systems of domination. Even so, few researchers have taken the difficult (and inherently geographical) step to show how resistance is written into the landscape, how it changes environments, and alters local ecologies.

Karl Zimmerer's "Wetland Production and Smallholder Persistence: Agricultural Change in a Highland Peruvian Region" (Zimmerer 1991) is unusual in this regard, and is in many ways an interesting model for political ecology realized too infrequently: a road less taken, between physical and social research. In this dense article Zimmerer engages some of the more traditional questions of peasant studies (Chapter 3): Why do smallholding peasants persist in the Andes when both Marxian and conservative neo-classical economic theory say that they should not? Turning detailed attention to the agronomy and history of the Colquepata district of highland Peru, Zimmerer shows not only physically *how* they do it, but also politically *why*.

The answer to the first question (how?) requires agronomic assessment. Zimmerer reveals that the peasant producers of the region manage, maintain, and produce an unusual wetland environment – drained-*wachu* agriculture – through carefully coordinated collective action. Exploring the physical dynamics of agrarian soils and drainage, he describes a sophisticated and elaborate network of canals, laid out in a herringbone pattern, that properly inundate and drain high-elevation potato fields. He also tracks the complex labor scheduling required to coordinate many difficult tasks, including weeding, plowing, and planting. This work shows the trained eye of someone at home both in physical and social science.

To answer the second question (why?) requires a detailed understanding of the region's political history. As an area that has strained to maintain its autonomy through successive periods, spanning the colonial and independence eras, Colquepata has always been a "region of resistance" where free communities have struggled over land and labor rights with surrounding hacienda estates and state elites. Resisting the urban bias of the country's leadership, and the skewed subsidies that disadvantaged the highlands, Colquepata peasants expanded and coordinated *wachu* agriculture and commercialized their own wetland production to persist and thrive.

This synthesis of agronomy and history, canals and land rights, uprisings and markets, reflects a background that is somewhat rare. Zimmerer's eclectic experience includes a BS in physics and biology, time spent at the Land Institute in Kansas, together with a Sauerian training at Berkeley with significant exposure to biogeography during his PhD. Political ecology commonly makes claims to this kind of interdisciplinarity, but learning to do it is more difficult. Zimmerer's work shows the substantial benefits of that kind of training.

of nature, therefore, and its legacy in contemporary forestry, *chipko* is an expression of local women's efforts to restore ecosystems, protect the reproductive power of the environment, and reassert nature. Challenging masculinist commercial exploitation of nature, *chipko* was a feminist movement.

The brutality and general truth of these sorts of assertions have been documented elsewhere. Differential gendered rights and obligations in environmental systems are nearly universal, and changes in institutions, markets, and technologies that impinge on environmental systems do tend to impact men and women quite differently (Chapter 9). The account, especially when told in the articulate urgency of Shiva's voice, certainly seems plausible.

Ram Guha's parallel account, *The Unquiet Woods* (Guha 1989), is written in an equally urgent voice. This similarly prominent book insists that *chipko* is one in a long line of peasant movements, sparked by the dissolution of a traditional and less-stratified Himalayan social structure, divided by the emergence of colonial and independence-era commercialism and land rents. Striking out in defense of their "moral economy," *chipko* peasants sought to reclaim their traditional systems of production and agitate against globalizing forestry.

Like Shiva's, this account seems to ring some bells, reminding us of conclusions based on field-oriented political ecological research, for example, Wilson's Sangre de Cristo residents (Chapter 9), Kull's fire-wielding herders and farmers (Chapter 8), and Hecht and Cockburn's Amazonian producers (Chapter 7). Though the account contradicts that of Shiva, it appears equally compelling at first glance.

Mythical movements and the risks of romance But as charged by observers like Rangan (2000), Mawdsley (1998), and others, the movement about which Guha and Shiva are arguing is a *myth*. Moreover, they charge, the efforts of Guha and Shiva to characterize this movement and propagate their own romantic accounts are ironically complicit in derailing the demands of those local people they claim to represent and defend.

This, according to Rangan, is because both Shiva's and Guha's accounts, no matter how politically critical, actually draw from similarly inherited romantic tropes and stories about the Himalayas. Specifically, the "ecological harmony" posited by Shiva and the "social harmony" offered by Guha, those states of society/nature to which these authors assert *chipko* is seeking a return, are old chestnuts of colonial story-telling. Himalayan "backwardness" and "ecological stability" are both fantasies. The Garhwal region has been integrated into regional and global exchanges since the precolonial period, and its "peasants" have by no means been an unstratified and market-isolated population. The environment, likewise, has historically been in a constant state of upheaval and change, with erosion, mass wasting, and forest cover loss, all part of a geologically active region. The local actions that have come to be known as *chipko* were not, and could not have been, efforts to restore something that never existed.

Rather than a struggle to keep trees from being cut down and defend ecological ethnic identity, the *chipko* actions of 1973 were geared towards protesting against the state forest department's allotment of forest cutting to large timber merchants rather than to local woodcutters, whose industrial contract allotments were abandoned in favor of a new concessionary system. The struggle, therefore, was not to maintain traditional preservation of forests, but instead to locally capture the flow of value coming from a highly politicized forest environment. Rather than rejecting modernization, markets, or even the state, Garhwal activists sought revenue from forest exploitation, but *on their own terms*. This central fact, so important to local people, is incompatible with the romances of *chipko* narratives, and is therefore forgotten (Rangan 2000).

Evaluating the Thesis

This branch of political ecology is one that borders most closely on the social, rather than biophysical, sciences. With its concerns about the character of social organization, the emergence of cultural networks, and the upheavals in political systems, it addresses the fields of political science, history, sociology, and anthropology. Its contributions to these fields are several, because it re-centers explanation of social and political change onto livelihoods, the most immediately biophysical moments of daily life. Research in this area argues for, and empirically supports, the fundamental ways that abstract human experiences and social processes like identity, ethnicity, and political agency are grounded in the most common material things, like fertilizers, drinking water, and trees; people make an identity as they make a living.

Of course, there are some conceptual risks inherent in any notion of livelihood movements. As the cases of the Andes, Skull Valley, and Garhwal suggest, the primitive essentialisms associated with "ecological ethnicity" may lead to romantic shortcuts rather than rigorous analysis. These pitfalls should not distract us from the fundamental challenges posed by people acting to secure their own social ecologies.

Making politics by making a living

Most political ecologists agree on certain general normative claims. People should be allowed to retain and maintain their own self-determination and control their own labor. Communities should be allowed to build collective institutions, redistribute and share risk, and maintain the dignity of the least fortunate. Ecologies should be maintained with an eye towards medium- and long-term human use, while attending to their inherent values and diversity. But in fact, most political ecological research tends to focus on the forces that lead to *the destruction of these very possibilities*. Processes of marginalization and degradation, conservation efforts gone haywire, and divided ecological politics usually point to what has gone wrong.

The value of livelihood movement research, therefore, is that it transcends this work, to show how ecologies are viewed, produced, and defended by local people. That these people don't always do things the way outsiders would like – using chemicals, negotiating for waste sites, cutting trees – is not a problem, indeed it is exactly the point. A postcolonial political ecology has to admit that it doesn't have all the answers, and that knowledge close to the ground is as legitimate a form of science as that of academic observers.

And the degree to which indigenous, traditional, and marginal people embrace elements of development, finding new ways to make an identity while they make a living, is all to the good. If chemical fertilizers can help keep *quichua* people on the land, and allow them to articulate their own traditional identity, does it matter that chemical fertilizers are not "traditional"? As Bebbington observes: "People encounter development from their mundane, daily concerns to build and improve their livelihoods, to build places they enjoy being in, to give meaning to their lives through these livelihoods and places, and to maintain, and, as far as possible, to extend the degree to which they can exercise control over their conditions of existence" (Bebbington 2000, p. 513).

The risk of primitive romances and essentialisms

At the same time, there remains a perilous risk of romantic essentialism in this work that leads not only to inaccurate assessments of local political ecology, but also serves to hamstring the very activities that are locally under way. As Rangan warns us, the motivations of local people in Garhwal – who sought industrial concessions – were radically re-presented by observers into "ecological movements" and acts of "ecological ethnicity" by "ecosystem people." In so arguing, expectations about local demands were set by people *outside* the region, however well-meaning. The apparently radical claims of Shiva and Guha, in their strategic essentialisms, only reinforce the colonial nature of power relations between Garhwal and the rest of India and the world. And in truth, much of the work in this area of political ecology has been marred with just this sort of romance.

The cure for such colonial romance, however, is not a retreat from political ecology. Rather it means doing more and better political ecological analysis. The only way in which Shiva and Guha could produce essentialist narratives of *chipko*, after all, is by sidelining voices, local political economies, specific livelihood practices, careful analysis of environmental history, and local conflicts on the ground; by *not* doing political ecology, in other words.

The reality of dissent

Even given the risk of romanticizing such communities and pigeonholing such movements, the fact remains that political ecological actions at the local scale often represent dissent. Local actions (whether or not they could be called "movements" of one kind or another) provide an alternative vision of politics and decision-making; by accepting some forms of modernization (fertilizer, nuclear waste, or industrial tree harvesting) while rejecting others, and doing so on their own terms, these communities also present an active face, challenging the homogenizing and exploiting forces of globalization.

In this sense, the very idea of "community," so often romantically used by outsiders to characterize local polities, can be strategically useful. While such groups may not actually be organic "communities," they can certainly represent themselves in that way to provide a united front, provoke sympathy, claim collective property, and muster their identity against the forces arrayed against them (Li 1996). These movements show that while it is impossible to "opt-out" of engagement with a globalizing world, it is by no means impossible to set some of the terms of engagement.

They have also shown that of all the strategies available to people as they adapt to their environment (e.g., agricultural technology, redistributive community networks, systems of exchange), political protest and demands placed on the state remain important. Indian federations in Ecuador have successfully secured local autonomy and control not by rejecting the state, but rather by insisting that the state enforce land reform. An unromantic accounting of livelihood struggles, which takes seriously the opportunities of modern technology, the importance of the state, and the inevitability of ongoing adaptation, remains a worthy research agenda.

In the Field: Pastoral Polities in Rajasthan

My own attempts to understand these relationships – between community, identity, livelihoods, and technology – center on the livelihood struggles of the *raika* community of southern Pali, in Rajasthan. This is a community of animal raisers who have historically been market-oriented, integrated with cropping, and whose livelihood is typified by annual migration (Kohler-Rollefson 1994). *Raika* households keep permanent homes in mixed-caste agricultural villages from which they take their herds of camels, goats, and sheep on migration during the dry season. The frequency of migration varies depending upon access to local pasturage (Prasad 1994), labor and capital (Agrawal 1999), and markets (Kavoori 1999).

Despite the high and growing demand for animals, animal products, and meat, the *raika*'s situation is somewhat tenuous. Pasturage continues to decline in the region, removing key resources for survival. The political position of *raika*, who are non-locals wherever they go on migration, is also difficult. As outsiders in a densely peopled, land-scarce country, they can be treated with hostility and violence.

At the same time, the *raika* are well organized in their own advocacy. The way in which they articulate their identity, organize their activities, and seize the opportunities offered by contemporary markets and technologies tells us a lot about the way agricultural change will occur around the world in the next few decades. Observing and working with a local non-governmental organization in Pali for several years, the Lok hit Pashu Palak Sansthan (LPPS or organization for the welfare of animal raisers), I was able to informally consider a few questions typical of queries in the political ecology of livelihoods: How do *raika* assert their social and political power? How does identity and livelihood articulate with modernization and technological innovation?

A study of this sort requires participant observation and a lot of hanging around listening. By watching *raika* interact with other producer groups, listening to their conversations with veterinary scientists and NGO workers, and tracking the stories of their political actions in the last few years, a picture of a livelihood movement emerges. These observations suggest some answers to the questions phrased above. *Raika* assert their social and political power by allying with other agrarian communities with whom they are linked through ecology of nutrient exchanges. When necessary, they call upon their extended social networks and family relations to demonstrate a show of force. They have embraced modern veterinary technology where it has proven useful, but continue to assert traditional knowledge, specifically as a political challenge to state veterinary technicians who are often hostile to the community. My experiences in this interaction, however, show the often-ambivalent politics of research on social movements.

Agrarian alliances and traditional technology as resistance

Watching the daily interactions of *raika* herders, it is clear where their ecological and political commitments lie. Every day their animals are penned on land belonging to other individuals, families, and groups. This ongoing ecological exchange (nutrients in crop waste for nutrients in dung) builds economic relationships that extend to other areas of exchange. Many non-*raika* families buy and sell their animals through the *raika* or loan their herds during the migration season, for example. As a result, the herders have strong regional ties to landholding groups. As a caste with long historical connections to the elite *rajput* community, *raika* also continue to cultivate alliances with landlords or local royalty. The political and ecological linkages between the *raika* and other communities are densely interwoven.

But the emergent political power of the *raika* does not depend solely on their centrality to nutrient and capital flows. Over the last few years, the level of organization amongst the *raika* has expanded significantly. When the community is challenged by the state, either through new forest codes or changes in rules governing access, they have organized, occasionally calling on their extensive networks in a show of force. In 2001, when a Supreme Court order closed local forests to grazing, for example, the community organized a rally of unprecedented size, bringing hundreds of *raika* to a small town in Pali, blocking traffic, and challenging local constabulary. Simultaneously, *raika* elders and representatives sat with officials to hammer out a compromise that would allow animals access to their traditional range within the state forest reserve. This exercise was one that depended both on intimidation and on a notion of livelihood-based collective rights. It proved extremely effective, and sections of the forest were opened immediately after the rally.

Clearly, the already traditionally strong caste identity of the *raika* has been mobilized to advocate its ecological interests. Though the organization of caste identity as political force is by no means unprecedented (Jeffrey 2001), the emergence of an ecological politics of caste is notable and certainly reflects some form of "ecological ethnicity."

At the same time, *raika* ecological identity has begun to extend to the less confrontational though equally political area of indigenous science. As a herding caste, the community maintains a large, collective, ethno-veterinary knowledge system, including natural remedies for sick animals as well as medical procedures with a long history of success (Geerlings 2001). Such knowledge is increasingly lost as state extension authorities promulgate professionalized animal care. These state vets avow an explicit hostility towards traditional *raika* medicine and take it as their responsibility to educate locals and eliminate traditional medicine, especially "obscure" practices like the branding of animals to heal internal disorders.

The emergence and articulation of these traditional practices and technologies, however, has accelerated in recent years. This is in part because the LPPS has championed the cause of these practices and knowledges. But it is also because an increasingly well organized and represented *raika* community has associated their community identity and politics with ecological practices. As predicted under a general thesis of "ecological ethnicity," therefore, an explicitly politicized ecology has become an organizing theme for a successful social movement, one that challenges state and professional authorities.

Research amongst the *raika* has allowed me a glimpse into how people organize to defend their interests and engage ecological and technological change as a political act. *Raika* politics tie together nutrients, medicines, agrarian assets, enclosures, migration, and land rights in a complex network.

Ambivalence, research, and ethics

But this research is itself a political act, since my formal interaction with the *raika* organization is one of instrumental exchange. I have produced advocacy statements, reports, and analyses, most of which support *raika* claims: that their practices are sustainable, their participation in the local economy is crucial, and their claims to land rights are defensible. That these claims are borne out by my analyses, at least so far, has made the relationship a useful one, for the NGO and for the *raika* movement as a whole, in admittedly a very limited way. In exchange, the LPPS and the *raika* have been forthcoming in allowing me access and introduction to important events and people.

My interactions with the community therefore raise some difficult questions about sympathies, advocacy, and ethics in fieldwork. These difficult questions are in part a result of my own ambivalence about *raika* politics.

There is no question that the herders' role in the function of the regional ecological system, through nutrient transport and deposition, is crucial to the productivity of forests and fields. Recent challenges to their traditional grazing rights are inherently unfair and problematic. As a middle-caste community, however, with quite a bit of political clout, and good connections in high political circles, these herders are not the most marginal community in the region by any means. Tribals (or *adivasis*), leather-workers, and day laborers, on the other hand, are far less organized than the *raika*, and their demands (often to reduce grazing in some forest areas) receive little time and attention owing to the emergent power of the *raika* polity. *Adivasi* (indigenous or tribal) communities, for example, whose back-breaking labor is a hidden engine of economic growth, are not represented by any NGO, and at times the interests of these two groups are divergent, if not contradictory. To the degree that *raika* politics sideline or contradict those of even more marginal groups, *raika* advocacy can and must be questioned on ethical grounds.

Of course, my tutelage in Rajasthani custom and politics has been almost entirely under the *raika* and members of the LPPS, to whom I owe a profound personal debt. My personal ties to the people and the area run deep.

There is no simple answer to this problem, though some possibilities are suggested. The political economic system that pits herders against laborers in the first place is one that is in itself dysfunctional and deserves ongoing challenge. The prospects for herder–laborer alliances, though remote, present an intriguing possibility for the future.

In the meantime, however, this "stickiness" that causes livelihoods, advocacy, and research to cling together presents a puzzle for appropriate and progressive work. Because it is embroiled inevitably in the social movements that stir there, research in Rajasthan is for me an ambivalent practice, therefore. It is one for which I feel a great passion and that allows me to pursue questions of endless interest, but also it forces me to ask who gains and who loses from research, and to think hard about how one ought to act in a political ecological landscape.

Part IV

Where to Now?

In which several criticisms of political ecology are aired and we consider some of the asymmetries in the field. An alternative path forward is offered, some new research paths are introduced, and the very high stakes of all this work are briefly reviewed.

11

Where to Now?

"Against Political Ecology"?

In a 1999 essay, Andrew Vayda and Bradley Walters made an argument "against political ecology," insisting that the move away from apolitical human ecologies of the past, while refreshing, had gone too far, producing an analytically weak and dogmatic research trajectory where political economic forces always determine ecological outcomes. Political ecologists, they charged, know the answer before they start to do their research. Instead researchers should pursue "event ecology," where single environmental events are explained inductively, and less politically, in an expanding set of causes and effects.

Only three years earlier, Richard Peet and Michael Watts offered an inverse criticism. They suggested that, rather than a clear and coherent theory to account for environmental change, political ecology offered haphazard, contextual, and *ad hoc* accounts following "chains of explanation" with no coherent privileging of central driving and systemic tendencies. Instead researchers should pursue a "liberation ecology," where a more political theory of political ecology might better direct coherent normative explanations of social and environmental change, using materialist conceptions of consciousness, poststructuralist theories of discourse, and some form of environmental determinism.

It would be easy to get the impression, if one were to simply read criticisms and not actual research in the field, that political ecology can't seem to get anything right. Either it is hedged-in by dogmatic theory or it has too many loose accounts of causation. Either it is too little concerned with environmental impacts, or too little concerned with the relative and constructed nature of environmental process; by trying to be all things to all people, perhaps political ecology has failed altogether.

Too much theory or too little?

In reviewing *Land Degradation and Society*, Peet and Watts conclude that political ecology offers only:

> an extremely diluted diffuse and on occasion volunteeristic series of explanations. Degradation can arise under falling, rising, or stable population pressures, under an upswing or downswing in the rural economy, under labor surplus or labor shortage; in sum under virtually any set of conditions . . . Political ecology is radically pluralist and largely without politics or an explicit sensitivity to class interest and social struggle. (Peet and Watts 1996b, p. 8)

Such a description seems alien to the literature I surveyed in the preparation of this volume. Whether in the record of the conservation state as an expression of colonial/postcolonial order (Chapter 8) or in the class and gender processes that define and redefine control of productive resources (Chapter 9), the persistence of a few key patterns and forces, even amidst diverse regional differences, is impressive.

Indeed, when united, these patterns form a coherent, if somewhat eclectic, theory of political ecology. Consider the theoretical claims, concepts, and conclusions that are demonstrated in research around each of the four theses summarized previously.

That land degradation:

- is a regionalized phenomenon conditioned by trans-regional patterns of accumulation and declining producer margins
- is mediated by the adaptation of land managers and the variable reversibility and multiple equilibria of environmental system states

That global conservation efforts:

- are ordered around the increasing governmentality and internalization of state rule
- tend towards the territorialization of conservation space despite the non-territorial character of social and ecological processes
- depend on the construction of "wilderness" where human populations have often been active agents of environmental change in the past
- present an aggressive challenge to historically important local social ecologies

That environmental conflicts:

- result from environmental development practices that are conditioned by classed, gendered, raced imaginaries
- fall along extant faultlines of regional social stratification that determine differential environmental access and responsibility
- embody classed and gendered struggles over highly malleable property institutions

That emerging environmental movements:

- are born of the differential risk and ecological injustice that develops from patterns of uneven development
- develop from challenges to traditional ecological economies and so lead to resistance centered around producer livelihoods
- rarely follow simple patterns of resistance to "development" and "modernization" articulated by romantic observers, since the terms of resistance are set on the anti-colonial margins

In light of these, it would be difficult to defend the claim that there "is no serious attempt" on the part of political ecologists "at treating the means by which control and access of resources or property rights are defined, negotiated, and contested within the political arenas of the household, the workplace, and the state" (Peet and Watts 1996b, pp. 8–9).

At the same time, political ecologists continue to hedge their bets before predicting anything so bold as a single set of structural forces under which land degradation must happen. The literature suggests complex networks that organize over time to produce new environments, each contextually quite different. Consider, for example, the complex case of St Vincent surveyed in Chapter 7. Clearly, global markets set the terms of nutrient extraction from regional soil, but the politicized local responses and innovations that result create highly variable outcomes. As Rangan explains, unlike its apolitical counterparts, theory in political ecology recognizes human/non-human relationships to be linked through dynamics that may yield unpredictable consequences (Rangan 2000, p. 63). In light of this kind of robust contextualized explanation (which leaves political ecologists open to the kind of criticisms launched by Peet and Watts), Vayda and Walters's claims are all the more unconvincing. It is certainly clear that there is no single, overriding, or dogmatic set of inevitable conclusions in political ecology, despite some claims to the contrary.

Denunciations versus asymmetries

These kinds of critiques really represent a form of mutual denunciation. Peet and Watts remain admirably skeptical of concepts like "event ecology" because they smack of *ad hoc* and apolitical science, long demonstrated to be a dead end. Vayda and Walters justifiably wonder where the environment has gone, as well as the surprise, in political ecological explanation. But even as these two positions denounce one another, they are forwarded on a very thin and purposive reading of the political ecological literature. What I hope this book has demonstrated is that this field is hard to characterize in any such blanket criticisms. While I am sure we can find samples of research that fit the contradictory profiles of Peet and Watts and Vayda and Walters, they poorly capture such a dynamic enterprise.

More than this, these critiques are, in many ways, false issues, in that they point to inevitable tensions in *all forms of explanation*, none of which are unique to political ecology. A single emancipatory theory of socio-environmental degradation, as called for by Peet and Watts, must inevitably collide with Vayda and Walters's call for an inductive event ecology. Vayda and Walters call for a refocusing on the material, while Peet and Watts call for a refocusing on the constructed. Vayda and Walters insist political explanation has gone too far, while Peet and Watts argue that it has not gone far enough. Each represents a denunciation of the other, but neither may have much bearing on the issues facing political ecology.

Having said this, the impatience that Peet and Watts and Vayda and Walters express in their criticisms does suggest several problems within the explanations dominant in political ecology. These are threefold. First, it is indeed difficult to reconcile a serious effort to understand the objective conditions of ecology, especially as a determinant biophysical explanation for social events or conditions, with our increasingly clear understanding of the constructed character of the categories we use to describe and assess that ecology. Second, it is difficult to see how the lessons and theoretical insights of traditional political ecology can be applied to processes away from third-world agrarian environments, even while we intuitively know that first-world cities and other spaces are enmeshed in the same processes. Third, it is hard to imagine an explanatory framework that takes seriously both the highly localized conditions of production as well as the power and impact of non-local players (states and firms) as well as those of non-human agents, without retreating to a crude global/local hierarchy on the one hand, or surrendering to determinism on the other.

Three Calls for Symmetry

I would argue that these problems are less about "too little" or "too much" theory, ecology, or anything else. Nor are they the result of anyone going "too far" in one direction or another, materialist, constructivist, or otherwise. Instead, I would argue that these weaknesses are a result of asymmetries in explanation. By opening the focal length of political ecology's lens to capture symmetrical processes, many of the apparent theoretical conflicts over explanation may yet disappear.

From destruction to production

The efforts of traditional political ecology have been directed as reactions to apolitical ecologies. As such, research has often focused on demonstrating that the causes of environmental "problems" (defined by ministries, media, and other powerful agents) were not always what they appeared to be. These problems, for example soil erosion, were assumed to exist in an unproblematic way, only their explanation was challenged.

As both better environmental science and genealogical approaches were applied to these problems, however, two parallel discoveries were made. First, the environment turned out to be more complex and variable than was previously known. This made a simple focus on explaining land degradation difficult, since the biophysical phenomenon of degradation became harder to define, measure, and predict, at least in any simple way. At the same time, historical examinations of conservation science were beginning to suggest that the very apolitical notion of "degradation" is itself a highly relative and power-laden concept (Demeritt 1994).

The implications of this happy convergence of thinking are twofold. First, it demonstrates that while tensions between scientific ecology (measuring degradation) and constructivism (defining degradation) are perhaps inevitable, they need not restrict cooperative and mutual exploration of social/environmental phenomena. "Why has the environment changed?" is a question inevitably intertwined with "How are the terms of change defined and by whom?"

Second, this increasing ambivalence towards prognostications of environmental degradation further suggests that a switch from metaphors of environmental *destruction* (or *construction*) to those of *production* would benefit all parties to the dispute. Following Neil Smith's (1996) essay on the subject, I would suggest simply that the environments around us, including and especially those composed of non-humans, are clearly produced. Forests are produced as much as factories, polar ice sheets as much as reservoirs, Yellowstone's wilderness as much as a toxic dump. That human beings are by no means the only players in the production of these spaces makes them no less artificial (in the sense of "created"). Indeed, as political ecologists continually emphasize, the environment is not a malleable thing outside of human beings, or a tablet on which to write history, but instead a produced set of relationships that include people, who, more radically, are themselves produced.

The case material reviewed in this volume seems to support such a proposition. Indeed, research in political ecology, whether by accident or by design, seems to meet the challenges laid down by Smith in his essay, simultaneously expressing: "the inevitability and creativity of the social relationship with nature; the very real project of domination embodied in the capitalist mode of production; the differentiated relationship with nature according to gender, class, race, sexual preference; the implausibility of autonomous nature; and a strong response to the almost instinctive romanticism which pervades most treatments of nature" (Smith 1996, p. 49).

This does not mean a retreat, however, from normative environmental struggles against undesirable outcomes: lost species, ugly life spaces, toxic landscapes. Just because all environments are produced does not mean all environments are inevitable, desirable, just, or sustainable. It simply represents a renunciation that there is a social/environmental state to which we can and should ever *return*. This does not undermine environmental struggles; it simply suggests we approach them with a new language. Indeed, this conception helps us to symmetrically imagine human and non-human processes in the landscape, surrendering a position of "mastery" over the non-human world; this being, after all, the ultimate goal of mainstream environmentalism (a movement Smith unfairly chastises).

From peasants to producers

A second tension in political ecological explanation is that between typical research in third-world environments amidst primary producers and emergent work in first-world environments amongst very different populations, including professionals, consumers, and other players in global political ecology. At first blush, it would certainly seem as though the tools and lessons from one location poorly transport to another. What does the "moral economy of the peasant," with its inherent orientation towards calculation of cropping risks, tell us about someone driving an SUV in Peoria, after all, or about a delegate to a conference on global warming? This conundrum has occupied the attention of several scholars recently, with some calling for regional approaches to the problem (North American political ecology, West African political ecology, etc.), rather than assuming first world/third world binaries (Walker 2003). What I would suggest, however, is that while political ecological concepts *seem* ill suited to different worlds, these differences might well dissolve if their frameworks were directly applied to differing and apparently alien circumstances (McCarthy 2002).

Consider, for example, the case of the American lawn, a consumer ecology with serious implications for human health and ecosystem integrity because of the high levels of inputs required in its maintenance. An explanation of the lawn, usually viewed as a unique historical artifact of American regional culture and therefore fundamentally different from crops in Africa or forests in India, has typically proceeded in an apolitical cultural-historical fashion. Where did the aesthetic come from? How was it introduced? (Jenkins 1994).

But by symmetrically imagining the lawn to be, on the contrary, exactly like third world food crops, political ecological doors are flung open. Consider: the lawn produces instrumental capital value for its cultivator in property values (Robbins et al. 2001); the lawn represents shared community capital and is enforced by cooperative/coercive neighborhood moral economies (Robbins and Sharp 2003); the lawn becomes a chemical treadmill for its owner since green revolutionary inputs are increasingly required to maintain its form and value (Robbins and Birkenholtz 2003); the lawn is a sink for global corporate chemical firms desperately seeking new markets amidst declining margins (Robbins and Sharp 2003). How does this differ from the complex and embedded decision context of a wheat cultivator in Rajasthan?

Analysis of systems of ecological production should therefore be extended to all manner of players and actors, treating them all as *producers*. This means doing detailed ethnographic analysis of soil science laboratories, ministry offices, and housing and urban development meetings, since lab technicians, ministers, and bureaucrats are all producers, enmeshed in moral economies, with structured production incentives, embedded in local knowledge systems, cognizant of the classed, gendered, and casted divisions in access and power as they produce landscapes.

From chains to networks

An equally vexing problem for political ecologists is the habit of explanation born of Vayda's progressive contextualization and Blaikie and Brookfield's chain of explanation. By always following explanation "upwards" from produced environments, through producers, and on to increasing scales of interaction (typically the community, the state, and the global economy), a conceptual hierarchy of power and causal force is imposed on political ecological problems that is empirically unfounded, and perhaps politically undesirable. Producers control landscape outcomes, their

Box 11.1 Michael Dove and the study of foresters, bureaucrats, scientists, and other indigenous people

If political ecology teaches us anything, it is that forestry can be a violent, accumulative, and ecologically problematic business. But what do we really know about the people who enact forestry programs, or about the way they imagine forests, farmers, soils, or anything else for that matter? Despite decades of research in political ecology, these questions seem mysteriously unanswered. As anthropologist Michael Dove notes: "The acknowledgment that the interests of the farmer must be reckoned with if forestry development is to succeed, while once a radical idea, is now widely accepted in forestry development. Yet development impasses and failures in the forestry sector persist, in part because one player remains to be recognized: the national forest services and their foresters" (Dove 1994, p. 333).

Dove further argues that the study of peasants rather than extension agents, and a focus on farmers instead of project planning rooms, is a habit of thinking inherited from traditional ethnography, and its focus on the exotic Other. Therefore, even where critical scholarship seeks to redress problems and advocate the rights of producers it remains focused on members of producer communities – farmers, herders, fishers – rather than examining state agencies and civil servants who create the conditions in which other people toil. This represents disproportionate attention to the "study of indigenous movements and NGOs rather than government ministries and to the study of local organizations of resistance rather than central organizations of oppression" (Dove 1999, p. 240).

As Dove insists, however, this problem is "a predisposition, but not an incapacitation. In the past decade or two, we have seen ample evidence of the capacity of ethnographers to study not just distant and marginal communities but close and powerful national and transnational institutions" (Dove 1999, p. 239).

Returning to our first question then. As it turns out, careful exploration of the social life of foresters reveals that they are often poor people, in marginal economic positions, with peculiar local ecological knowledges, situated within confined fields of agency by socio-economic structures of environmental control and power, all within conservation discourses promulgated by distant elites (Robbins 2000). In other words, they are a lot like the rest of us, except that their official context constrains their imagination of the world in a specific way. These official constructions of reality often create a barrier to more democratic and sustainable practices, but they are amenable to change, if they are well understood.

Michael Dove has opened a box for political ecologists that has been closed too long. A new generation of scholars will have to take up this challenge, don their pith helmets, and enter the thickets of ministry offices, forestry nurseries, water quality laboratories: the realms of many intriguing indigenous peoples.

behavior is prefigured by community dynamics, set within state policies, controlled by trade agreements, all within a vast system vaguely described as "global capitalism." The ascent in scales also imposes a "chain of command" where players at distantly removed scales (peasants, states, the World Trade Organization) have little interaction.

But using this approach, how do we understand the relationship of varying producers of nature – those on the farm, in the lab, and in the office – as they mutually create the landscape while mutually coercing one another? Moreover, how do we understand the role of non-human actors – soil microbes, animals, and trees – as they participate in this production, acting either in a cooperative fashion (some trees grow well in agroforestry plots and increase production) or an uncooperative one (some trees favor human disturbance habitats but further invade and decrease production)? The chain of explanation is a poor conceptual tool to manage such linkages and relationships. A shift away from this way of thinking towards a comparative anatomy of *networks* is therefore a more viable mode of explanation.

Networks organize and are organized by a range of human and non-human actors, through systems of accumulation, extraction, investment, growth, reproduction, exchange, cooperation, and coercion. While diverse, each network is by no means unique. Common patterns of exploitation and environmental change reflect common network morphologies and common processes. By explicating networks, therefore, we come to a better understanding of recurrent socio-natural situations, especially undesirable ones. Rather than manipulating or waiting for changes in global political economy to trickle down a chain, a network allows us a range of places for progressive political action and normative change.

The Hybridity Thesis

None of the revelations I have suggested above are altogether new, even if they are somewhat fresh to formal political ecology. A focus on production of nature rather than destruction/construction follows the call of social studies of science and technology (Latour and Woolgar 1979; Latour 1987; Porter 1995; Latour 1999; see Hess 1997 for a comprehensive overview). A focus on networks reflects ongoing evolution of explanation in the social sciences, but especially in critical development studies (Bebbington and Batterbury 2001). An increasing interest in non-peasant producers, laboratories, and bureaucrats is well established in feminist approaches to science (Haraway 1989; Merchant 1989; Harding 1990; Keller 1995) and in critical analyses of policy science (Salter 1988; Jasanoff 1990). Together this work paints a picture of a produced world, where politics hold sway, but which involves global migrants, both human and non-human, who produce and consume landscapes and knowledge, remaking the world as they go. This is nowhere more evident than in the circulation of non-human biological agents through the global political economy, including the plants, animals, and microbes, which, in the words of Spencer and Whatmore (2001), create as they continuously collide with human affairs, "the circumstances and anxieties of today's world, characterized as it is by a widespread sense of a massive and irreversible socialization of the bio-physical world, whether purposefully through biodiversity conservation programmes or unintentionally through global climate change processes" (p. 140). Clearly a shift towards production of nature, producer politics, and networks of interaction – a "more-than-human geography" – is on the agenda throughout the social sciences, and political ecologists would benefit from the innovation it offers (Whatmore 2003).

At the same time, however, research in cultural and political ecology brings many things to this agenda that are *clearly lacking*. First, political ecological work has historical depth, which reminds us that such collisions, between plants and people, animals and crops, or soils and research institutions, are neither new nor unique to first world practices of genetic engineering. Indeed it is that very modern conceit – that society is at last triumphant, having mastered the boundary walls between nature and culture – that disintegrates under the scrutiny of political ecology surveyed previously in this volume.

Political ecology also has a highly flexible focal length, which crosses scale and frames its analysis as easily on global institutions and scientific labs, a focus typical of much work in critical science studies, as on peasants, hunters, and homemakers, constituencies which are sometimes overlooked. Similarly, political ecology brings a problem orientation to these issues that is immediate, practical, and designed to show flaws and propose alternatives to real policy measures, a much-vaunted goal in critical environmental studies (Castree 2002).

Finally, cultural and political ecologists, especially the former but to no small degree the latter, are experienced in examining biophysical processes or at least talking to biophysical scientists. This is so much the case that when Whatmore (2003) advocated hybrid research by stating that "perhaps it is too late to repair the gulf that has opened up between human and physical skills that once permitted conversations between cultural geography and biogeography" (p. 139), one would need only consider the wealth of skills and practices of political ecologists described in this volume to imagine a retort.

The resulting approach indicates a new argument, which for lack of a better name I call here *the hybridity thesis*. It suggests certain tendencies and trends in the collision of human and non-human nature and paves the way for new research. While this thesis is as yet perhaps too underdeveloped to paraphrase here, its abstract outlines are suggested.

The hybridity thesis: the ecological characteristics of non-human nature and its objects (dung, climate, maize, lawn grass, bacteria, goats, and tropical soils) impinge upon the political world of human struggles. Yet as these characteristics and agents are altered in their interactions with humans, whether purposefully or unintentionally, they assume new roles, set new terms, and take on new importance. People, institutions, communities, and nations assemble and participate in the networks created in this interaction, leveraging power and influence, just as non-human organisms and communities do. In recent history, powerful modern institutions and individuals (environmental ministries, multinational corporations, corrupt foresters) have gained undue and disproportionate power by explicitly attempting to divide and police the boundaries between human and non-human nature, even while allying themselves and building new connections to the non-human world, leading to unintended consequences and pernicious results. In the process, resistance emerges from traditional, alternative, and progressive human/non-human alliances marginalized by such efforts (usually along lines of gender, class, and race).

Political ecologies of success

Far from a revolutionary break with the way political ecologists are thinking, this network approach is essentially a distillation of *de facto* practice in the field. It is perhaps best exemplified by several recent works on peasant adaptation to changing markets and local environments, which are surprising since they show success stories, rather than the usual disasters that are more commonly the stuff of political ecology.

Lawrence Grossman's *Political Ecology of Bananas* (Grossman 1998) and Tom Bassett's *Peasant Cotton Revolution in West Africa* (Bassett 2001) both startlingly conclude that the development of commodity production for the global market amongst small producers, which typically leads to social and environmental trouble, has proved successful. In both cases – Grossman's St Vincent and Bassett's northern Cote D'Ivoire – peasant production of cash crops has led to relatively stable incomes, stable ecologies, and stable communities, despite predictions to the contrary. Why should this be so?

In answering, Bassett and Grossman both draw attention to the specific networks formed between (1) biophysical conditions, processes, and agents, (2) producers and producer communities, and (3) state actors and policies. In Cote D'Ivoire rural peasants did not abandon their social networks and modes of exchange in their transition to intensive cotton production. They did, however, reconfigure their cropping rotations and crop ecologies as they simultaneously began to negotiate new gendered rules for labor and expand their use of new systems of credit and technology. Most significantly, local producers began to make demands on the state, organizing strikes and forcing new relationships with state pricing boards. New alliances between state actors and producer groups set the stage for a surprisingly progressive outcome in a global environment otherwise typified by marginalization (Bassett 2001).

In St Vincent the growth of contract farming resulted in the maintenance of peasant communities, rather than their dissolution. Here, the specific ecological conditions of banana production made the extension of an industrial model of processing difficult. Producer cooperatives, formed in and through the state, spread risk and costs through new, wider channels, and allowed locally specific practices and technologies to thrive. Like the Ivoirian case, St Vincent defies the predicted decline of the producer, predicated on the decline of the state, in a context of supranational markets and institutions. This is because a specific configuration of ecological elements, producers, producer groups, and state agents was formed to resist and adapt to change, a causal process that poorly resembles a vertical chain.

Both of these networked outcomes may indeed be quite fragile. Major changes in global crop prices or the institutional autonomy of these small countries in the face of GATT (General Agreement on Tariffs and Trade) and the World Trade Organization, for example, would force serious changes in either situation. The robustness of these networks may make them adaptive to changing conditions, however. So too, the common configurations of these networks may provide a model for other producer polities around the world, as they similarly seek to retain their rights, their land, and their dignity. Consider, for example, the concerted and successful pressure exerted by poorer cotton-producing nations in challenging what they saw to be unfair trade restrictions during the October 2003 Cancun negotiations of the WTO; these inversions of "free trade" in commodities challenge relationships between the environmental periphery (like Cote D'Ivoire and Brazil) and the established core (the United States and the European Union).

New substantive research mandates

Like political ecologies of success, there are obviously countless substantive areas of research that remain unexamined in political ecology, including the political ecology of Eastern European transition, the political ecology of the drug trade, and the political ecology of trash, waste management, and garbage subsistence. But the conceptual re-imaginings that follow from the hybridity thesis draw attention to three specific areas of concern that need far more attention than they currently receive: human population dynamics, genetically modified landscapes, and the urban environment.

Population is too important to be left to the Malthusians The primary legacy of Malthusian explanation for political ecology is that researchers in the field are constantly on guard for its logics and constantly unteaching it in classrooms and development institutions. Massive demographic shifts are occurring around the world, however, that are worthy of critical attention. Consider some parts of Latin America, where birth rates are falling precipitously through changing fertility behavior, or AIDS-driven depopulation in many parts of the world. What do these mean for household labor, the politics of local service provision, and the environmental power of differing age cohorts? What new ecologies will be formed in such demographic and technological shifts? What new farms, factories, and markets may emerge in the aftermath of population decline?

Conversely, many tribal communities, who were on the brink of extinction only a few decades ago, are achieving tremendous population growth and expanding in an explicitly political effort to secure and occupy traditional territories, many of which are conservation areas. What might this portend for future environmental dynamics and relative political power in tribal areas and the states within which they are nested? The aversion political ecologists feel for some population theory must not keep them from understanding the political ecology of demography.

Genetic modification won't go away In the last few years, Monsanto's Bt Bollgard cotton was released legally in South Africa. It was also planted in Zimbabwe without permission and has applications pending in many southern African nations. United States companies have further introduced the plant in Zambia despite state resistance (Kuyek 2002).

Bt plants, inserted with the genes that make the soil microbes of *Bacillus thuringiensis* toxic to insects with the aim of creating a plant that will naturally resist common pests, represent a novel influence on politics, economy, and ecology, like other transgenically modified cultivars. Yet unknown is the degree to which these plants may cross-pollinate with pernicious weeds. So too, it is unclear how quickly these new cultivars are replacing local agricultural diversity. Nor are the local peasant economics of these species clear, or whether the plant's introduction may lead to consolidation or instead to opportunities for small producers. State efforts to keep such plants out of their borders, moreover, are challenged both by illegal introduction and incidental contact and pollination.

The ramifications of these introductions for landscapes, local power relations, and global economic exchanges are tremendous. Critical environmental research has so far been slow to empirically engage the question, however, and little work has been done by political ecologists to examine transgenic landscapes in detail on the ground. If the effects of these changes, which seem now to have a lot of momentum, are to be tracked and the forces at work in their introduction challenged, political ecologists will have to start now.

Cities are political ecologies While we often think of the densest flows of nutrients, elements, and energy to be those of tropical rainforests and diverse savannas, they are certainly at their most complex in urban environments. City streets, gardens, golf courses, kitchen sinks, and garages are all teeming with life, connected and regulated through systems of power and fixed through investments of capital. Cities offer many of the same complex conditions as a rural town, moreover: environmental conservation remains a mechanism for control; marginalization of communities is tied to the production of new, and often undesirable, ecologies; political conflicts are commonly articulated as ecological ones, owing to differential human access and responsibility in nature; social movements emerge from the daily business of making a living.

A political ecology of the city can expand beyond simply identifying the unequal distribution of risks and environmental "bads" to explain how these urban ecologies are produced and why these ecological networks look the way they do. Tracing flows and clusters of garbage, trees, energy, runoff, and disease through built urban space, and examining governance of green spaces, both public and private, should be all the easier in a political ecology informed by a notion of the produced character of nature. Agnostic to whether or not a forested "wilderness" or a suburb is more natural, political ecology might integrate critical theories of urban growth, decay, investment, and control with ecosystem analysis of daily life.

Against "Against Political Ecology": Retaining both Theory and Surprise

Obviously, there is too much important work to do to surrender the field to any new form of "ecology" that purports to explain a complex world without a broad range of critical tools. Even so, many of the old habits of political ecological explanation will have to be transcended if the field is to move forward conceptually. Political ecologists will have to seek simultaneously the theoretical coherence suggested by Peet and Watts, while opening their methods to the empirical surprise suggested by Vayda and Walters. It will also mean wrestling with a bold new range of substantive topics. The next generation of research stands to make real strides in critically explaining environmental problems and facilitating the environmental successes to come.

In the Meantime . . .

In the meantime, however, forests in Asia are being cut down to line the pockets of already wealthy officials and global timber traders. Simultaneously, monocultural tree plantations have displaced traditional tribal forest and grazing lands from Latin America to South Africa.

Each United States citizen consumes 1,600 liters of gasoline every year, with obvious implications for global warming. Petroleum exporting countries like Nigeria remain some of the world's poorest.

Migrant workers exposed to agricultural chemical pesticides experience daily bouts of weakness, fatigue, nausea, muscle pains, and cramps, while living with a grossly heightened risk of leukemia. Voluntary applications of similar organophosphate pesticides on middle-class lawns put children and ambient ecosystems at risk. Together these uses fuel a multi-billion-dollar global industry.

During the last century 90 percent of global agricultural crop diversity has been lost, even while pests and diseases are mutating and expanding at an accelerating pace. Even so, introgressed transgenic maize has been introduced into central Mexico, the original center and origin of native maize landrace diversification, risking serious genetic decline.

The best-selling books that address these issues, on the other hand, insist that the disparity of nations is a product of the shape of continents (Diamond 1997) and that environmental crises are statistical fictions (Lomborg 2001).

If political ecology has taught us anything, it is that we can do better than that. We can do better than that.

References

Acheson, J. M. (1997) "The Politics of Managing the Maine Lobster Industry: 1860 to the Present." *Human Ecology* **25**: 3–27.

Addams, J. (1910) *Twenty Years at Hull-House*. New York, MacMillan Company.

Adler, R. W., J. C. Landman and D. M. Cameron (1993) *The Clean Water Act: 20 Years Later*. Washington DC, Island Press.

Agee, J. K. (1993) *Fire Ecology of Pacific Northwest Forests*. Washington DC, Island Press.

Agrawal, A. (1995) "Dismantling the Divide between Indigenous and Scientific Knowledge." *Development and Change* **26** (3): 413–39.

Agrawal, A. (1999) *Greener Pastures: Politics, Markets, and Community among a Migrant Pastoral People*. Durham, Duke University Press.

Akram-Lodhi, A. H. (1992) "Peasants and Hegemony in the Work of James C. Scott." *Peasant Studies* **19** (3–4): 179–201.

Alland, A. (1975) "Adaptation." *Annual Review of Anthropology* **4**: 59–73.

Allen, B. and R. Crittenden (1987) "Degradation and Pre-capitalist Political Economy: The Case of the New Guinea Highlands." In *Land Degradation and Society*, edited by P. Blaikie and H. Brookfield. London, Routledge: 145–56.

Althusser, L. and É. Balibar (1970) *Reading Capital*. New York, Pantheon Books.

Alveres, C. (1993) Science. In *The Development Dictionary*, edited by W. Sachs. London, Zed Books: 219–32.

Anderson, D. and R. Grove (1987) "Introduction: The Scramble for Eden: Past, Present and Future in African Conservation." In *Conservation in Africa: People, Policies and Practice*, edited by D. Anderson and R. Grove. Cambridge, Cambridge University Press: 1–12.

Anderson, J. E. and R. S. Inouye (2001) "Landscape-scale Changes in Plant Species Abundance and Biodiversity of a Sagebrush Steppe over 45 Years." *Ecological Monographs* **71** (4): 531–56.

Aston, T. and C. Philpin, eds. (1985) *The Brenner Debate: Agrarian Class Struggle and Economic Development in Pre-industrial Europe*. Cambridge, Cambridge University Press.

Bailes, K. E., ed. (1985) *Environmental History: Critical Issues in Comparative Perspective*. New York, University Press of America.

Bakker, K. (1999) "The Politics of Hydropower: Developing the Mekong." *Political Geography* **18** (2): 209–32.

Barham, B. and O. T. Coomes (1996) *Prosperity's Promise: The Amazon Rubber Boom and Distorted Economic Development*. Boulder, Westview Press.

Barnes, T. J. and J. S. Duncan (1992) "Introduction: Writing Worlds." In *Writing Worlds: Discourse, Text, and Metaphor in the Representation of Landscape*, edited by T. J. Barnes and J. S. Duncan. New York, Routledge: 1–17.

Barth, F. (1956) "Ecological Relationships of Ethnic Groups in Swat, North Pakistan." *American Anthropologist* **58**: 1079–89.

Bassett, T. J. (1988) "The Political Ecology of Peasant–Herder Conflicts in the Northern Ivory Coast." *Annals of the Association of American Geographers* **78** (3): 453–72.

Bassett, T. J. (2001) *The Peasant Cotton Revolution in West Africa: Côte d'Ivoire, 1880–1995*. Cambridge, Cambridge University Press.

Batterbury, S. (2001) "Landscapes of Diversity: A Local Political Ecology of Livelihood Diversification in South-western Niger." *Ecumene* **8** (4): 437–64.

Batterbury, S. and A. Warren (2001) "The African Sahel 25 Years after the Great Drought: Assessing Progress and Moving towards New Agendas and Approaches." *Global Environmental Change* **11** (1): 1–8.

Batterbury, S., T. Forsyth and K. Thomson (1997) "Environmental Transformations in Developing Countries: Hybrid Research and Democratic Policy." *The Geographical Journal* **163** (2): 126–32.

Bayliss-Smith, T. P. (1982) *The Ecology of Agricultural Systems*. Cambridge, Cambridge University Press.

Bebbington, A. (1993) "Modernization from Below: An Alternative Indigenous Development?" *Economic Geography* **69** (3): 274–92.

Bebbington, A. (1996) "Movements, Modernizations, and Markets: Indigenous Organizations and Agrarian Strategies in Ecuador." In *Liberation Ecologies*, edited by R. Peet and M. Watts. New York, Routledge: 86–109.

Bebbington, A. (2000) "Reencountering Development: Livelihood Transitions and Place Transformations in the Andes." *Annals of the Association of American Geographers* **90** (3): 495–520.

Bebbington, A. and S. Batterbury (2001) "Transnational Livelihood and Landscape: Political Ecologies of Globalization." *Ecumene* **8** (4): 369–80.

Been, V. and F. Gupta (1997) "Coming to the Nuisance or Going to the Barrios? A Longitudinal Analysis of Environmental Justice Claims." *Ecology Law Quarterly* **24** (2): 1–56.

Behnke, R. H. and I. Scoones, eds. (1993) *Range Ecology at Disequilibrium: New Models of Natural Variability and Pastoral Adaptation in African Savannas*. London, Overseas Development Institute.

Bennett, J. (1969) *Northern Plainsmen: Adaptive Strategy and Agrarian Life*. New York, Aldine.

Benton, T., ed. (1996) *The Greening of Marxism*. New York, Guilford Press.

Berkes, F. (1999) *Sacred Ecology: Traditional Ecological Knowledge and Resource Management*. Philadelphia, Taylor and Francis.

Blaikie, P. (1985) *The Political Economy of Soil Erosion in Developing Countries*. New York, Longman Scientific and Technical.

Blaikie, P. (1999) "A Review of Political Ecology." *Zeitschrift fur Wirtschaftsgeographie* **43**: 131–47.

Blaikie, P. and H. Brookfield (1987) *Land Degradation and Society*. London and New York, Methuen and Co. Ltd.

Blaikie, P., J. Cameron and D. Seddon (1980) *Nepal in Crisis: Growth and Stagnation at the Periphery*. Oxford, Clarendon Press.

Blaut, J. M. (1999) "Environmentalism and Eurocentrism." *Geographical Review* **89** (3): 391–408.

Blaut, J. M. (2000) *Eight Eurocentric Historians*. New York, Guilford Press.

Blumler, M. A. (1998) "Biogeography of Land-use Impacts in the Near East." In *Nature's Geography: New Lessons for Conservation in Developing Countries*, edited by K. S. Zimmerer and K. R. Young. Madison, WI, University of Wisconsin Press: 215–36.

Bonnard, P. and S. Scherr (1994) "Within Gender Differences in Tree Management: Is Gender Distinction a Reliable Concept?" *Agroforestry Systems* **25** (2): 71–93.

Botkin, D. B. (1990) *Discordant Harmonies: A New Ecology for the Twenty-first Century*. New York, Oxford University Press.

Bowen, J. R. (1986) "The War of Words: Agrarian Change in Southeast Asia." *Peasant Studies* **14** (1): 55–62.

Brass, T. (2000) "Moral Economists, Subalterns, New Social Movements and the (Re-) Emergence of a (Post-) Modernized (Middle) Peasant." In *Mapping Subaltern Studies and the Postcolonial*, edited by V. Chaturvedi. London, Verso Press: 127–62.

Braudel, F. (1982) *Civilization and Capitalism* vol. 2: *The Wheels of Commerce*. New York, Harper and Row Publishers.

Bridge, G. (2000) "The Social Regulation of Resource Access and Environmental Impact: Production, Nature, and Contradiction in the US Copper Industry." *Geoforum* **31**: 237–56.

Brodt, S. (1999) "Interactions of Formal and Informal Knowledge Systems in Village-based Tree Management in India." *Agriculture and Human Values* **16** (4): 355–63.

Brodt, S. B. (2001) "A Systems Perspective on the Conservation and Erosion of Indigenous Agricultural Knowledge in Central India." *Human Ecology* **29** (1): 99–120.

Brokensha, D., D. M. Warren and O. Werner, eds. (1980) *Indigenous Knowledge Systems and Development*. Washington, DC, University Press of America.

Bromley, D. (1992) *Making the Commons Work: Theory, Practice, and Policy*. San Francisco, Institute for Contemporary Studies.

Brookfield, H. (1962) "Local Study and Comparative Method: An Example from Central New Guinea." *Annals of the Association of American Geographers* **52** (3): 242–54.

Brookfield, H. C. and P. Brown (1963) *Struggle for Land: Agriculture and Group Territories among the Chimbu of the New Guinea Highlands*. New York, Oxford University Press.

Brush, S. B. and D. Stabinsky, eds. (1996) *Valuing Local Knowledge*. Washington DC, Island Press.

Bryant, R. (1999) "A Political Ecology for Developing Countries." *Zeitschrift fur Wirtschaftsgeographie* **43** (3–4): 148–57.

Bryant, R. L. (1996) "The Greening of Burma: Political Rhetoric or Sustainable Development." *Pacific Affairs* **69** (3): 341–59.

Bryant, R. L. (2002) "Non-governmental Organizations and Governmentality: 'Consuming' Biodiversity and Indigenous People in the Philippines." *Political Studies* **50** (2): 268–92.

Bryant, R. L. and S. Bailey (1997) *Third World Political Ecology*. New York, Routledge.

Bullard, R. D. (1990) *Dumping in Dixie*. Boulder (CO), Westview.

Burger, J. and M. Gochfeld (1998) "The Tragedy of the Commons 30 Years Later." *Environment* **40** (10): 4.

Burton, I., R. W. Kates and G. White (1993) *The Environment as Hazard*. New York, The Guilford Press.

Butzer, K. (1976) *Early Hydraulic Civilization in Egypt: A Case Study in Cultural Ecology*. Chicago, University of Chicago Press.

Butzer, K. (1989) "Cultural Ecology." In *Geography in America*, edited by C. J. Wilmott and G. L. Gaile. New York, Merrill.

Butzer, K. (1992) "The Americas before and after 1492: Current Geographical Research." *Annals of the Association of American Geographers* **82** (3): 345–68.

Campbell, S. G., R. B. Hanna, M. Flug and J. F. Scott (2001) "Modeling Klamath River System Operations for Quantity and Quality." *Journal of Water Resources Planning and Management – ASCE* **127** (5): 284–94.

Capel, H. (1994) "The Imperial Dream: Geography and the Spanish Empire in the Nineteenth Century." In *Geography and Empire*, edited by A. Godlewska and N. Smith. Oxford, Blackwell: 58–73.

Carney, J. (1993) "Converting the Wetlands, Engendering the Environment: The Intersection of Gender with Agrarian Change in the Gambia." *Economic Geography* **69** (4): 329–48.

Carney, J. (2001) *Black Rice: The African Origins of Rice Cultivation in the Americas*. Cambridge, Harvard University Press.

Carney, J. and M. Watts (1991) "Disciplining Women? Rice, Mechanization, and the Evolution of Mandinka Gender Relations in Senegambia." *Signs* **16** (4): 651–81.

Carson, R. (1962) *Silent Spring*. New York, Houghton Mifflin.

Carty, R. K. and M. Eagles (1998) "The Political Ecology of Local Party Organization: The Case of Canada." *Political Geography* **17** (5): 589–609.

Cashman, K. (1991) "Systems of Knowledge as Systems of Domination: The Limitations of Established Meaning." *Agriculture and Human Values* **8** (1/2): 49–58.

Cassidy, C. (1980) "Benign Neglect and Toddler Malnutrition." In *Social and Biological Predictors of Nutritional Status, Physical Growth, and Neurological Development*, edited by L. S. Greene and F. E. Johnston. New York, Academic Press: 109–33.

Castree, N. (2002) "Environmental Issues: From Policy to Political Economy." *Progress in Human Geography* **26** (3): 357–65.

Center for Research on Economic Development: University of Michigan (1977) "Marketing, Price Policy and Storage of Food Grains in the Sahel: A Survey." Ann Arbor, MI, Center for Research on Economic Development, University of Michigan.

Chaturvedi, V. (2000) *Mapping Subaltern Studies and the Postcolonial*. London, Verso.

Chayanov, A. V. (1986) *The Theory of Peasant Economy*. Madison, University of Wisconsin Press.

Chief Wildlife Warden Kumbhalgarh Wildlife Sanctuary (1996) Management Plan: Kumbhalgarh Wildlife Sanctuary 1996–1997 to 2000–2001. Udaipur, Rajasthan Forest Service.

Ciriacy-Wantrup, S. V. and R. C. Bishop (1975) "Common Property as a Concept in Natural Resources Policy." *Natural Resources Journal* **15**: 713–27.

Cleary, M. and P. Eaton (1996) *Tradition and Reform: Land Tenure and Rural Development in South-east Asia*. Oxford, Oxford University Press.

Clements, H. (1983) *Alfred Russel Wallace: Biologist and Social Reformer*. London, Hutchinson.

Cockburn, A. and J. Ridgeway, eds. (1979) *Political Ecology*. New York, New York Times Book Company.

Cohen, S. (1999) "Promoting Eden: Tree Planting as the Environmental Panacea." *Ecumene* **6** (4): 424–46.

Collett, D. (1987) "Pastoralists and Wildlife: Image and Reality in Kenya Massailand." In *Conservation in Africa: People, Policies and Practice*, edited by D. Anderson and R. Grove. Cambridge, Cambridge University Press: 129–48.

Commoner, B. (1988) "The Environment." In *Crossroads: Environmental Priorities for the Future*, edited by P. Borelli. Washington DC, Island Press: 121–69.

Commons, J. R. (1990) *Institutional Economics*. New Brunswick, Transaction Publishers.

Conklin, H. C. (1954) "An Ethnoecological Approach to Shifting Agriculture." *New York Academy of Sciences, Transactions* **17** (2): 133–42.

Conklin, H. C. (1968) "Some Aspects of Ethnographic Research in Ifugao." *Transactions of the New York Academy of Sciences* **30**: 99–121.

Cribb, R., ed. (1990) *The Indonesian Killings of 1965–1966: Studies from Java and Bali*. Clayton, Victoria (Australia), Centre of Southeast Asian Studies, Monash University.

Cronon, W. (1983) *Changes in the Land*. New York, Hill and Wang.

Cronon, W. (1992) *Nature's Metropolis: Chicago and the Great West*. New York, W. W. Norton and Co.

Cronon, W. (1995) "The Trouble with Wilderness, or, Getting Back to the Wrong Nature." In *Uncommon Ground: Rethinking the Human Place in Nature*, edited by W. Cronon. New York, W. W. Norton and Co.: 69–90.

Crosby, A. W. (1986) *Ecological Imperialism: The Biological Expansion of Europe, 900–1900*. Cambridge, Cambridge University Press.

Crotty, R. M. (1980) *Cattle, Economics, and Development*. Slough, Commonwealth Agricutural Bureau.

Cruz, M. C., C. A. Meyer, R. Reppeto and R. Woodward (1992) *Population Growth, Poverty, and Environmental Stress: Frontier Migration in the Phillipines and Costa Rica*. Washington DC, World Resources Institute.

Cutter, S. (1995) "Race, Class, and Environmental Justice." *Progress in Human Geography* **19**: 107–18.

Dallmeier, F., ed. (1992) *MAB Digest 11: Long-term Monitoring of Biological Diversity in Tropical Forest Areas: Methods for Establishment and Inventory of Permanent Plots*. Paris, UNESCO.

Daniels, R. and T. J. Bassett (2002) "The Spaces of Conservation and Development around Lake Nakuru National Park, Kenya." *Professional Geographer* **54** (4): 481–90.

Darnovsky, M. (1992) "Stories Less Told: Histories of US Environmentalism." *Socialist Review* **22** (4): 11–56.

Darwin, C. (1860) *On the Origin of Species by Means of Natural Selection, or, The Preservation of Favoured Races in the Struggle for Life*. London, J. Murray.

Das Gupta, M. (1987) "Selective Discrimination Against Female Children in Rural Punjab." *Indian Population Development Review* **13**: 77.

Das Gupta, M. (1995) "Life Course Perspectives on Women's Autonomy and Health Outcomes." *American Anthropologist* **97** (3): 481–91.

Davis, A. and J. R. Wagner (2003) "Who Knows? On the Importance of Identifying 'Experts' when Researching Local Ecological Knowledge." *Human Ecology* **31** (3): 463–89.

Davis, M. (2001) *Late Victorian Holocausts: El Niño Famines and the Making of the Third World*. London, Verso.

de Janvry, A. (1981) *The Agrarian Question and Reformism in Latin America*. Baltimore, Johns Hopkins University Press.

Demeritt, D. (1994) "Ecology, Objectivity, and Critique in Writings on Nature and Human Societies." *Journal of Historical Geography* **20**: 22–37.

Demeritt, D. (1998) "Science, Social Constructivism and Nature." In *Remaking Reality: Nature at the Millennium*, edited by B. Braun and N. Castree. New York, Routledge: 173–93.

Demeritt, D. (2001) "The Construction of Global Warming and the Politics of Science." *Annals of the Association of American Geographers* **91** (2): 307–37.

Denevan, W. (1973) "Development and the Imminent Demise of the Amazon Rain Forest." *Professional Geographer* **25** (2): 130–35.

Denevan, W. M. (1992) "The Pristine Myth: The Landscape of the Americas in 1492." *Annals of the Association of American Geographers* **82** (3): 369–85.

Denevan, W. M. (2001) *Cultivated Landscapes of Native Amazonia and the Andes: Triumph over the Soil*. Oxford, Oxford University Press.

Diamond, J. (1997) *Guns, Germs, and Steel: The Fates of Human Societies*. New York, W. W. Norton.

Doolittle, W. E. (2000) *Cultivated Landscapes of Native North America*. Oxford, Oxford University Press.

Dove, M. (1983) "Theories of Swidden Agriculture and the Political Economy of Ignorance." *Agroforestry Systems* **1**: 85–99.

Dove, M. (1994) "The Existential Status of the Pakistani Farmer: Studying Official Constructions of Social Reality." *Ethnology* **33** (4): 331–51.

Dove, M. (1995) "The Theory of Social Forestry Intervention: The State of the Art in Asia." *Agroforestry Systems* **30**: 315–40.

Dove, M. (1999) "Writing for, Versus about, the Ethnographic Other: Issues of Engagement and Reflexivity in Working with a Tribal NGO in Indonesia." *Identities* **6** (2–3): 225–53.
Dove, M. (2000) "The Life Cycle of Indigenous Knowledge, and the Case of Natural Rubber Production." In *Indigenous Environmental Knowledge and its Transformations: Critical Anthropological Perspectives*, edited by R. Ellen, P. Parkes and A. Bicker. Amsterdam, Harwood Academic Publishers: 213–51.
Dove, M. R. and D. M. Kammen (2001) "Vernacular Models of Development: An Analysis of Indonesia under the 'New Order'." *World Development* **29** (4): 619–39.
Dumont, L. (1966) *Homo Hierarchicus: The Caste System and its Implications*. Chicago, University of Chicago Press.
Durrenberger, E. P. and N. Tannenbaum (1979) "A Reassessment of Chayanov and his Recent Critics." *Peasant Studies* **8** (1): 48–63.
Eastman, R., J. Mckendry and M. Fulk (1991) *Change and Time Series Analysis*. Geneva, UNITAR.
Eckholm, E. P. (1976) *Losing Ground: Environmental Stress and World Food Prospects*. New York, Norton.
Ehrlich, P. R. (1968) *The Population Bomb*. New York, Ballantine Books.
Ellen, R. (1982) *Environment, Subsistence and System: The Ecology of Small Scale Social Formations*. Cambridge, Cambridge University Press.
Ellen, R. (1998) "Palms and the Prototypicality of Trees: Some Questions Concerning Assumptions in the Comparative Study of Categories and Labels." In *The Social Life of Trees*, edited by L. Rival. Oxford, Berg.
Ellis, F. (1993) *Peasant Economics: Farm Households and Agrarian Development*. Cambridge, Cambridge University Press.
Emel, J. (1998) "Are you Man Enough, Big and Bad Enough? Wolf Eradication in the US." In *Animal Geographies: Place, Politics, and Identity in the Nature–Culture Borderlands*, edited by J. Wolch and J. Emel. London, Verso Press.
Emel, J. and R. Roberts (1995) "Institutional Form and its Effects on Environmental Change: The Case of Groundwater in the Southern High Plains." *Annals of the Association of American Geographers* **85** (4): 664–83.
Emel, J., R. Roberts and D. Sauri (1992) "Ideology, Property, and Groundwater Resources – An Exploration of Relations." *Political Geography* **11** (1): 37–54.
Emmanuel, A. (1972) *Unequal Exchange: A Study of the Imperialism of Trade*. New York, Monthly Review Press.
Enzensberger, H. M. (1996) "A Critique of Political Ecology." In *The Greening of Marxism*, edited by T. Benton. New York, Guilford Press: 17–49.
Erickson, J., D. Chapman and R. Johnny (1995) "Monitored Retrievable Storage of Spent Nuclear Fuel in Indian Country: Liability, Sovereignty, and Socioeconomics." *American Indian Law Review* **19**: 17–103.
Fairhead, J. and M. Leach (1995) "False Forest History, Complicit Social Analysis: Rethinking some West African Environmental Narratives." *World Development* **23** (6): 1023–35.
Fairhead, J. and M. Leach (1996) *Misreading the African Landscape: Society and Ecology in a Forest–Savanna Mosaic*. Cambridge, Cambridge University Press.
Fairhead, J. and M. Leach (1998) *Reframing Deforestation: Global Analysis and Local Realities: Studies in West Africa*. New York, Routledge.
Feeny, D., F. Berkes, B. J. McCay and J. M. Acheson (1990) "The Tragedy of the Commons – 22 Years Later." *Human Ecology* **18** (1): 1–19.
Foote, D. C. and B. Greer-Wooten (1968) "An Approach to Systems Analysis in Cultural Geography." *Professional Geographer* **20** (2): 86–91.
Forsyth, T. (1996) "Science, Myth, and Knowledge: Testing Himalayan Environmental Degradation in Thailand." *Geoforum* **27** (3): 375–92.

Forsyth, T. (2003) *Critical Political Ecology: The Politics of Environmental Science*. London, Routledge.

Fortmann, L. (1996) "Gendered Knowledge: Rights and Space in Two Zimbabwe Villages." In *Feminist Political Ecology*, edited by D. Rocheleau, B. Thomas-Slayter and E. Wangari. New York, Routledge.

Foster, J. B. (2000) *Marx's Ecology: Materialism and Nature*. New York, Monthly Review Press.

Foucault, M. (1971) *The Order of Things: An Archaeology of the Human Sciences*. New York, Pantheon Books.

Foucault, M. (1980) "Truth and Power." In *Power/Knowledge: Selected Interviews and other Writings 1972–1977*, edited by C. Gordon. New York, Pantheon: 109–33.

Foucault, M. (1991) "Governmentality." In *The Foucault Effect: Studies in Governmentality*, edited by G. Burchell, C. Gordon, and P. Miller. London, Harvester: 87–104.

Franke, R. and B. H. Chasin (1980) *Seeds of Famine: Ecological Destruction and the Development Dilemma in the Western Sahel*. Montclair, NJ, Allanheld, Osmun.

Franke, R. W. and B. H. Chasin (1979) "Peanuts, Peasants, Profits, and Pastoralists: The Social and Economic Background to Ecological Deterioration in Niger." *Peasant Studies* **8** (3): 1–30.

Freed, S. A. and R. S. Freed (1981) "Sacred Cows and Water Buffalo in India: The Use of Ethnography." *Current Anthropology* **22** (5): 483–90.

Fuhlendorf, S. D. and D. M. Engle (2001) "Restoring Heterogeneity on Rangelands: Ecosystem Management Based on Evolutionary Grazing Patterns." *BioScience* **51** (8): 625–32.

Gadgil, M. and R. Guha (1995) *Ecology and Equity: The Use and Abuse of Nature in Contemporary India*. London and New York, Routledge.

Gates, B. T. and A. B. Shteir, eds. (1997) *Natural Eloquence: Women Reinscribe Science*. Madison, University of Wisconsin Press.

Geerlings, E. (2001) "Sheep Husbandry and Healthcare System of the Raikas in South-central Rajasthan." Sadri (India), Local Livestock for Empowerment of Rural People (LIFE), League for Pastoral People.

Geertz, C. (1963) *Agricultural Involution: The Processes of Ecological Change in Indonesia*. Berkeley, University of California Press.

Gentry, A. H. (1986) "Species Richness and Floristic Composition of Choco Region Plant Communities." *Caldasia* **15**: 71–91.

George, S. (1990) "Agropastoral Equations in India: Intensification and Change of Mixed Farming Systems." In *The World of Pastoralism*, edited by J. G. Galaty and D. L. Johnson. New York, Guilford Press: 119–44.

Gilmartin, D. (1995) "Models of the Hydraulic Environment: Colonial Irrigation, State Power and Community in the Indus Basin." In *Nature Culture Imperialism: Essays on the Environmental History of South Asia*, edited by D. Arnold and R. Guha. Bombay, Oxford University Press: 210–36.

Goldman, M. (1991) "Cultivating Hot Peppers and Water Crisis in India's Desert: Toward a Theory of Understanding Ecological Crisis." *Bulletin of Concerned Asian Scholars* **23** (4): 19–29.

Goodman, D. and M. Redclift (1991) *Refashioning Nature: Food, Ecology, and Culture*. New York, Routledge.

Gordon, H. S. (1954) "The Economic Theory of a Common Property Resource: The Fishery." *Journal of Political Economy* **62**: 124–42.

Gould, S. J. (1996) *The Mismeasure of Man*. New York, W. W. Norton.

Gray, L. (1999) "Is Land being Degraded? A Multi-scale Investigation of Landscape Change in Southwestern Burkina Faso." *Land Degradation and Development* **10** (4): 329–43.

Greenberg, J. B. and T. K. Park (1994) "Political Ecology." *Journal of Political Ecology* **1**: 1–12.

Greenland, D. J., P. J. Gregory and P. H. Nye, eds. (1998) *Land Resources: On the Edge of the Malthusian Precipice?* New York, CAB International.

Grossman, L. (1977) "Man–Environment Relations in Anthropology and Geography." *Annals of the Association of American Geographers* **67** (1): 126–44.

Grossman, L. (1984) *Peasants, Subsistence Ecology, and Development in the Highlands of Papua New Guinea*. Princeton, Princeton University Press.

Grossman, L. (1998) *The Political Ecology of Bananas: Contract Farming, Peasants, and Agrarian Change in the Eastern Caribbean*. Chapel Hill, University of North Carolina Press.

Grossman, L. S. (1993) "The Political Ecology of Banana Exports and Local Food Production in St. Vincent, Eastern Caribbean." *Annals of the Association of American Geographers* **83** (2): 347–67.

Grove, R. H. (1990) "Colonial Conservation, Ecological Hegemony and Popular Resistance: Towards a Global Synthesis." In *Imperialism and the Natural World*, edited by J. M. MacKenzie. Manchester (UK), Manchester University Press: 15–50.

Guha, R. (1989) *The Unquiet Woods: Ecological Change and Peasant Resistance in the Himalaya*. New Delhi, Oxford University Press.

Gupta, A. (1998) *Postcolonial Developments: Agriculture in the Making of Modern India*. New Delhi, Oxford University Press.

Guyot, A. (1873) *Guyot's Physical Geography*. New York, American Book Company.

Hacking, I. (1999) *The Social Construction of What?* Cambridge, Harvard University Press.

Hanna, S. S., C. Folke and K.-G. Maler, eds. (1996) *Rights to Nature: Ecological, Economic, Cultural, and Political Principles of Institutions for the Environment*. Washington DC, Island Press.

Haraway, D. (1989) *Primate Visions: Gender, Race, and Nature in the World of Modern Science*. New York, Routledge.

Hardesty, D. (1975) "The Niche Concept: Suggestions for its Use in Human Ecology." *Human Ecology* **3** (2): 71–85.

Hardesty, D. (1977) *Ecological Anthropology*. New York, John Wiley and Sons.

Hardin, G. (1968) "The Tragedy of the Commons." *Science* **162**: 1243–8.

Harding, S. (1990) "Feminism, Science, and Anti-enlightenment Critiques." In *Feminism/Postmodernism*, edited by L. Nicholson. New York, Routledge: 83–106.

Harris, M. (1966) "The Cultural Ecology of India's Sacred Cattle." *Current Anthropology* **7** (1): 51–66.

Harrison, M. (1999) *Climates and Constitutions*. Oxford, Oxford University Press.

Harrison, Y. A. and C. M. Shackleton (1999) "Resilience of South African Communal Grazing Lands after the Removal of High Grazing Pressure." *Land Degradation and Development* **10** (3): 225–39.

Hart, G. (1991) "Engendering Everyday Resistance: Gender, Patronage and Production Politics in Rural Malaysia." *Journal of Peasant Studies* **19** (1): 93–121.

Harvey, D. (1996) *Justice, Nature, and the Geography of Difference*. Cambridge (US), Blackwell Publishers.

Hecht, S. and A. Cockburn (1989) *The Fate of the Forest: Developers, Destroyers and Defenders of the Amazon*. London, Verso.

Hecht, S. B., A. B. Anderson and P. May (1988) "The Subsidy from Nature: Shifting Cultivation, Successional Palm Forests and Rural Development." *Human Organization* **47** (1): 25–35.

Heffernan, M. J. (1994) "The Science of Empire: The French Geographical Movement and the Forms of French Imperialism, 1870–1920." In *Geography and Empire*, edited by A. Godlewska and N. Smith. Oxford, Blackwell: 92–114.

Hempel, L. C. (1996) *Environmental Governance: The Global Challenge*. Washington DC, Island Press.

Herrnstein, R. and C. Murray (1994) *The Bell Curve*. New York, Free Press.

Hess, D. J. (1997) *Science Studies: An Advanced Introduction*. New York, New York University Press.

Hindess, B. and P. Q. Hirst (1975) *Pre-capitalist Modes of Production*. London, Routledge and Kegan Paul.

Hocking, D., ed. (1993) *Trees for Drylands*. New Delhi, Oxford and IBH Publishing.

Homewood, K., E. F. Lambin, E. Coast et al. (2001) "Long-term Changes in Serengeti-Mara Wildebeest and Land Cover: Pastoralism, Population, or Policies?" *Proceedings of the National Academy of Sciences* **98**: 12544–9.

Honoré, A. M. (1961) "Ownership." In *Oxford Essays in Jurisprudence*, edited by A. G. Guest. Oxford, Clarendon Press: 107–47.

Humboldt, A. (1811) *Political Essay on the Kingdom of New Spain*. London, Longman, Hurst, Rees, Orme, and Brown.

Humboldt, A. (1852) *Personal Narrative of Travels to the Equinoctial Regions of America*. London, Henry G. Bohn.

Humboldt, A. (1858) *Cosmos*. New York, Harper and Brothers.

Hunt, R. C. (1988) "Size and the Structure of Authority in Canal Irrigation Systems." *Journal of Anthropological Research* **44** (4): 335–55.

Huntington, E. (1915) *Civilization and Climate*. New Haven, Yale University Press.

Huston, M. A. (1979) "A General Hypothesis of Species Diversity." *The American Naturalist* **113** (1): 81–101.

Huxley, T. H. (1896) *Evolution and Ethics, and other Essays*. New York, D. Appleton and Company.

Inden, R. (1990) *Imagining India*. Cambridge (MA), Blackwell Publishers.

Ishiyama, N. (2003) "Environmental Justice and American Indian Tribal Sovereignty: Case Study of a Land-use Conflict in Skull Valley, Utah." *Antipode* **35** (1): 119–39.

Jain, D. and N. Banerjee (1985) *Tyranny of the Household: Investigative Essays on Women's Work*. New Delhi, Shakti.

Jain, R. R. (1992) "Botanical and Phytogenic Resources of Rajasthan." In *Geographical Facets of Rajasthan*, edited by H. S. Sharma and M. L. Sharma. Chandranagar (India), Kuldeep Publications: 68–75.

Jarosz, L. (1993) "Defining and Explaining Tropical Deforestation – Shifting Cultivation and Population Growth in Colonial Madagascar (1896–1940)." *Economic Geography* **69** (4): 366–79.

Jasanoff, S. (1990) *The Fifth Branch: Science Advisers as Policymakers*. Cambridge, Cambridge University Press.

Jeanrenaud, S. (2002) "Changing People/Nature Representations in International Conservation Discourses." *Institute of Development Studies Bulletin* **33** (1): 111.

Jeffery, R. and B. Vira, eds. (2001) *Conflict and Cooperation in Participatory Natural Resource Management*. New York, Palgrave.

Jeffrey, C. (2001) " 'A Fist is Stronger than Five Fingers': Caste and Dominance in Rural North India." *Transactions of the Institute of British Geographers* **26** (2): 217–36.

Jenkins, V. S. (1994) *The Lawn: A History of an American Obsession*. Washington and London, Smithsonian Institute Press.

Jewitt, S. (1995) "Europe's 'Others'? Forestry Policy and Practices in Colonial and Post-colonial India." *Environment and Planning D: Society and Space* **13**: 67–90.

Jewitt, S. and S. Kumar (2000) "A Political Ecology of Forest Management." In *Political Ecology: Science, Myth, and Power*, edited by P. Stott and S. Sullivan. New York, Arnold: 91–113.

Jodha, N. S. (1985) "Population Growth and Decline of Common Property Resources in Rajasthan, India." *Population and Development Review* **11**: 247–64.

Jodha, N. S. (1987) "A Case Study of the Degradation of Common Property Resources in India." In *Land Degradation and Society*, edited by P. Blaikie and H. Brookfield. London, Routledge: 186–205.

Johnson, D. L. and L. A. Lewis (1995) *Land Degradation: Creation and Destruction*. Cambridge (MA), Blackwell Publishers.

Jokisch, B. (2002) "Migration and Agricultural Change: The Case of Smallholder Agriculture in Highland Ecuador." *Human Ecology* **30** (4): 523–50.

Jokisch, B. D. (1997) "From Labor Circulation to International Migration: The Case of South-central Ecuador." *Yearbook Conference of Latin Americanist Geographers* **23**: 63–75.

Kant, I. (1882) *Critique of Pure Reason*. London, George Bell and Sons.

Kavoori, P. S. (1999) *Pastoralism in Expansion: The Transhuming Herders of Western Rajasthan*. New Delhi, Oxford University Press.

Keller, E. F. (1995) "The Origin, History, and Politics of the Subject Called 'Gender and Science.'" In *Handbook of Science and Technology Studies*, edited by S. Jasanoff, G. E. Markle, J. C. Peterson and T. Pinch. London, Sage Publications: 80–94.

Kepe, T. and I. Scoones (1999) "Creating Grasslands: Social Institutions and Environmental Change in Mkambati Area, South Africa." *Human Ecology* **27** (1): 29–54.

Kitchin, R. and S. Frendschuh (2000) "Cognitive Mapping." In *Cognitive Mapping: Past Present and Future*, edited by R. Kitchin and S. Frendschuh. London, Routledge: 1–8.

Klooster, D. J. (2002) "Toward Adaptive Community Forest Management: Integrating Local Forest Knowledge with Scientific Forestry." *Economic Geography* **78** (1): 43–70.

Kohler-Rollefson, I. (1994) "Pastoralism in India from a Comparative Perspective: Some Comments." *Overseas Development Institute Pastoral Development Network* **36a**: 3–5.

Kropotkin, P. (1888) *Mutual Aid: A Factor in Evolution*. Boston, Porter Sargent Publishers.

Kropotkin, P. (1985) *Fields, Factories, and Workshops Tomorrow*. London, Freedom Press.

Kropotkin, P. (1987) *The State: Its Historic Role*. London, Freedom Press.

Kropotkin, P. (1990) *The Conquest of Bread*. Montreal, Black Rose Press.

Kuhlken, R. (1999) "Settin' the Woods on Fire: Rural Incendiarism as Protest." *Geographical Review* **89** (3): 343–63.

Kuhn, T. S. (1970) *The Structure of Scientific Revolutions*. Chicago, University of Chicago Press.

Kull, C. A. (1999) "Observations on Repressive Environmental Policies and Landscape Burning Strategies in Madagascar." *African Studies Quarterly* **3** (2): 1–7.

Kull, C. A. (2000) "Deforestation, Erosion, and Fire: Degradation Myths in the Environmental History of Madagascar." *Environment and History* **6** (4): 423–50.

Kull, C. A. (2002) "Empowering Pyromaniacs in Madagascar: Ideology and Legitimacy in Community-based Natural Resource Management." *Development and Change* **33** (1): 57–78.

Kumar, A. B., K. N. Tiwari, R. S. Dwivedi and D. Karunakar (1997) "Spectral Behavior and Spectral Separability of Eroded Land Using Multi-sensor Data." *Land Degradation and Development* **8** (1): 27–38.

Kumar, M. and M. M. Bhandari (1993) "Impact of Human Activities on the Pattern and Process of Sand Dune Vegetation in the Rajasthan Desert." *Desertification Bulletin* **22**: 45–54.

Kummer, D. M. (1992) *Deforestation in the Postwar Philippines*. Chicago, University of Chicago Press.

Kuyek, D. (2002) "Genetically Modified Crops in Africa: Implications for Small Farmers." Barcelona, Genetic Resources Action International.

Lal, R., D. O. Hansen, N. Uphoff and S. Slack, eds. (2002) *Food Security and Environmental Quality in the Developing World*. Boca Raton, Florida, CRC Press.

Laney, R. (2002) "Disaggregating Induced Intensification: A Case Study from Madagascar." *Annals of the Association of American Geographers* **92** (4): 702–26.

Latour, B. (1987) *Science in Action: How to Follow Scientists and Engineers through Society.* Cambridge, Harvard University Press.

Latour, B. (1993) *We have Never been Modern.* Cambridge, Harvard University Press.

Latour, B. (1998) "To Modernize or Ecologise? That is the Question." In *Remaking Reality: Nature at the Millenium*, edited by B. Braun and N. Castree. London and New York, Routledge: 221–42.

Latour, B. (1999) *Pandora's Hope: Essays on the Reality of Science Studies.* Cambridge, Harvard University Press.

Latour, B. and S. Woolgar (1979) *Laboratory Life: The Construction of Scientific Facts.* Princeton (NJ), Princeton University Press.

Leighly, J. (1965) "Introduction." In *Land and Life: A Selection of the Writings of Carl Ortwin Sauer*, edited by J. Leighly. Berkeley, University of California Press: 1–8.

Lemonick, M. D. and A. Dorfman (1994) "Too Few Fish in the Sea." *Time* **143** (14): 70.

Lenin, V. I. (1972a) *The Development of Capitalism in Russia.* Moscow, Progress Publishers.

Lenin, V. I. (1972b) "'Left Wing' Communism – an Infantile Disorder." In *Lenin: Collected Works*, vol. 31. Moscow, Progress Publishers: 17–118.

Lentz, D. L., ed. (2000) *Imperfect Balance: Landscape Transformations in the Precolumbian Americas.* New York, Columbia University Press.

Li, T. M. (1996) "Images of Community: Discourse and Strategy in Property Relations." *Development and Change* **27** (3): 501–27.

Liddle, J. and R. Joshi (1986) *Daughters of Independence: Gender, Caste, and Class in India.* New Brunswick (NJ), Rutgers University Press.

Livingstone, D. N. (1994) "Climate's Moral Economy: Science, Race and Place in Post-Darwinian British and American Geography." In *Geography and Empire*, edited by A. Godlewska and N. Smith. Oxford, Blackwell Publishers: 132–54.

Lockwood, J. A. and D. R. Lockwood (1993) "Catastrophe Theory: A Unified Paradigm for Rangeland Ecosystem Dynamics." *Journal of Range Management* **46** (4): 282–8.

Lodrick, D. O. (1994) "Rajasthan as a Region: Myth or Reality." In *The Idea of Rajasthan: Explorations in Regional Identity*, vol. 1, edited by K. Schomer, J. L. Erdman, D. O. Lodrick and L. I. Rudolph. New Delhi, Manohar and the American Institute of Indian Studies: 1–44.

Lomborg, B. (2001) *The Skeptical Environmentalist: Measuring the Real State of the World.* Cambridge, Cambridge University Press.

Lu, D., E. Moran and P. Mausel (2002) "Linking Amazonian Secondary Succession Forest Growth to Soil Properties." *Land Degradation and Development* **13** (4): 331–43.

Luke, T. W. (1999) "Eco-managerialism: Environmental Studies as a Power/Knowledge Formation." In *Living with Nature: Environmental Politics as Cultural Discourse*, edited by F. Fischer and M. A. Hajer. Oxford, Oxford University Press: 103–20.

Lykke, A. M. (2000) "Local Perceptions of Vegetation Change and Priorities for Conservation of Woody-savanna Vegetation in Senegal." *Journal of Environmental Management* **59** (2): 107–20.

McCarthy, J. (2002) "First World Political Ecology: Lessons from the Wise Use Movement." *Environment and Planning A* **34** (7): 1281–1302.

McCay, B. J. and J. M. Acheson, eds. (1987) *The Question of the Commons: The Culture and Ecology of Communal Resources.* Tucson, University of Arizona Press.

MacIntyre, A. A. (1987) "Why Pesticides Received Extensive Use in America: A Political Economy of Agricultural Pest Management to 1970." *Natural Resources Journal* **27**: 533–78.

McSweeney, K. (2004) (in press) "Approaching Rural Livelihoods through Systems of Exchange: The Dugout Canoe Trade in Central America's Mosquitia." *Annals of the Association of American Geographers.*

Malthus, T. R. (1992) *An Essay on the Principle of Population (Selected and Introduced by D. Winch)*. Cambridge, Cambridge University Press.

Mansfield, B. (2001) "Property Regime or Development Policy? Explaining Growth in the US Pacific Groundfish Fishery." *Professional Geography* **53** (3): 384–97.

Mansfield, B. (2003) "From Catfish to Organic Fish: Making Distinctions about Nature as Cultural Economic Practice." *Geoforum* **34**: 329–42.

Marco, G. J., R. M. Hollingsworth and W. Durham, eds. (1987) *Silent Spring Revisited.* Washington DC, American Chemical Society.

Marsh, G. P. (1898) *The Earth as Modified by Human Action.* New York, Charles Scribner's Sons.

Martinez-Alier, J. (2002) *The Environmentalism of the Poor: A Study of Ecological Conflicts and Valuation.* Cheltenham, Glos, Edward Elgar.

Marx, K. (1967a) *Capital*, vol. 1. New York, International Publishers.

Marx, K. (1967b) *Capital*, vol. 2. New York, International Publishers.

Mawdsley, E. (1998) "After Chipko: From Environment to Region in Uttaranchal." *Journal of Peasant Studies* **25** (4): 36–54.

Meadows, D. H., D. L. Meadows, J. Randers and W. W. I. Behrens (1972) *The Limits to Growth: A Report for the Club of Rome's Project on the Predicament of Mankind*, New York, Universe Books.

Meinzen-Dick, R. S., L. R. Brown, H. S. Feldstein and A. R. Quisumbing (1997) "Gender and Property Rights: Overview." *World Development* **25** (8): 1299–302.

Merchant, C. (1989) *Ecological Revolutions: Nature, Gender, and Science in New England.* Chapel Hill, University of North Carolina Press.

Messer, E. (1997) "Intra-household Allocation of Food and Health Care: Current Findings and Understandings – Introduction." *Social Science and Medicine* **44** (11): 1675–84.

Miller, V., M. Hallstein and S. Quass (1996) "Feminist Politics and Environmental Justice: Women's Community Activism in West Harlem, New York." In *Feminist Political Ecology: Global Issues and Local Experiences*, edited by D. Rocheleau, B. Thomas-Slayter and E. Wangari. New York, Routledge: 62–85.

Mitchell, D. (2000) *Cultural Geography: A Critical Introduction.* Oxford, Blackwell.

Mongia, P. (1997) "Introduction." In *Contemporary Postcolonial Theory*, edited by P. Mongia. London, Arnold: 1–18.

Moran, E. F. (1993) "Deforestation and Land Use in the Brazilian Amazon." *Human Ecology* **21** (1): 1–21.

Muldavin, J. S. S. (1996) "The Political Ecology of Agrarian Reform in China: The Case of Helongjiang Province." In *Liberation Ecologies: Environment, Development, and Social Movements*, edited by R. Peet and M. Watts. New York, Routledge: 227–59.

Murphy, R. F. (1981) "Julian Steward." In *Totems and Teachers: Perspectives on the History of Anthropology*, edited by S. Silverman. New York, Columbia University Press.

Murton, J. (1999) "Population Growth and Poverty in Machakos District, Kenya." *Geographical Journal* **165**: 37–46.

Mutersbaugh, T. (1999) "Bread or Chainsaws? Paths to Mobilizing Household Labor for Cooperative Rural Development in an Oaxacan Village (Mexico)." *Economic Geography* **75** (1): 43–58.

National Research Council (1986) *Proceedings of the Conference on Common Property Resource Management.* Washington DC, National Academy Press.

Netting, R. M. (1981) *Balancing on an Alp: Ecological Change and Continuity in a Swiss Mountain Community.* Cambridge, Cambridge University Press.

Netting, R. M. (1986) *Cultural Ecology.* Prospect Heights (IL), Waveland Press.

Netting, R. M. (1993) *Smallholders, Householders: Farm Families and the Ecology of Intensive, Sustainable Agriculture.* Stanford (CA), Stanford University Press.

Neumann, R. P. (1996) "Dukes, Earls, and Ersatz Edens: Aristocratic Nature Preservationists in Colonial Africa." *Environment and Planning D* **14** (1): 79–98.

Neumann, R. P. (1997) "Primitive Ideas: Protected Area Buffer Zones and the Politics of Land in Africa." *Development and Change* **28** (3): 559–82.

Neumann, R. P. (1998) *Imposing Wilderness: Struggles over Livelihood and Nature Preservation in Africa*. Berkeley, University of California Press.

Newman, R. P. (1992) *Owen Lattimore and the "Loss" of China*. Berkeley, University of California Press.

Nietschmann, B. (1972) "Hunting and Fishing Focus among the Miskito Indians, Eastern Nicaragua." *Human Ecology* **1** (1): 41–67.

Nietschmann, B. (1973) *Between Land and Water*. New York, Seminar Press.

Nietschmann, B. (1979) *Caribbean Edge: The Coming of Modern Times to Isolated People and Wildlife*. Indianapolis, Bobbs-Merrill.

Nietschmann, B. (1989) *The Unknown War: The Miskito Nation, Nicaragua, and the United States*. Lanham (MD), University Press of America.

Nietzsche, F. (1967) *On the Genealogy of Morals and Ecce Homo*. New York, Random House.

Oba, G., N. C. Stenseth and W. J. Lusigi (2000) "New Perspectives on Sustainable Grazing Management in Arid Zones of Sub-Saharan Africa." *BioScience* **50** (1): 35–51.

O'Connor, J. (1996) "The Second Contradiction of Capitalism." In *The Greening of Marxism*, edited by T. Benton. New York, Guilford Press: 197–221.

O'Faircheallaigh, C. (1998) "Resource Development and Inequality in Indigenous Societies." *World Development* **26** (3): 381–94.

Orlove, B. S. (1980) "Ecological Anthropology." *Annual Review of Anthropology* **9**: 235–73.

Ostrom, E. (1990) *Governing the Commons: The Evolution of Institutions for Collective Action*. Cambridge, Cambridge University Press.

Ostrom, E. (1992) *Crafting Institutions for Self-governing Irrigation Systems*. San Francisco, Institute for Contemporary Studies.

Ostrom, E., L. Schroeder and S. Wynne (1993) *Institutional Incentives and Sustainable Development*. Boulder, Westview Press.

Ovuka, M. (2000) "More People, More Erosion? Land Use, Soil Erosion and Soil Productivity in Murangaposa District." *Land Degradation and Development* **11** (2): 111–24.

Palsson, G. and A. Helgason (1995) "Figuring Fish and Measuring Men: The Individual Transferable Quota System in the Icelandic Cod Fishery." *Ocean and Coastal Management* **28** (1–3): 117–46.

Pandey, R. K., J. W. Maranville and T. W. Crawford (2001) "Agriculture Intensification and Ecologically Sustainable Land Use Systems in Niger: Transition from Traditional to Technologically Sound Practices." *Journal of Sustainable Agriculture* **19** (2): 5–24.

Parajuli, P. (1998) "Beyond Capitalized Nature: Ecological, Ethnicity as an Arena of Conflict in the Regime of Globalization." *Ecumene* **5** (2): 186–217.

Parayil, G. and F. Tong (1998) "Pasture-led to Logging-led Deforestation in the Brazilian Amazon: The Dynamics of Socio-environmental Change." *Global Environmental Change – Human and Policy Dimensions* **8** (1): 63–79.

Pastor, M., J. Sadd and J. Hipp (2001) "Which Came First? Toxic Facilities, Minority Move-in, and Environmental Justice." *Journal of Urban Affairs* **23** (1): 1–21.

Patterson, E. C. (1987) "Mary Fairfax Greig Somerville (1780–1872)." In *Women of Mathematics: A Bibliographic Sourcebook*, edited by L. Grinstein and P. Campbell: 208–16.

Peet, R. (1991) *Global Capitalism*. London, Routledge.

Peet, R. (1999) *Theories of Development*. New York, Guilford.

Peet, R. and M. Watts (1996a) *Liberation Ecologies: Environment, Development, Social Movements*. New York, Routledge.

Peet, R. and M. Watts (1996b) "Liberation Ecology: Development, Sustainability, and Environment in the Age of Market Triumphalism." In *Liberation Ecologies: Environment, Development, Social Movements*. New York, Routledge: 1–45.

Peluso, N. (1995) "Whose Woods are these? Counter-mapping Forest Territories in Kalimantan, Indonesia." *Antipode* **27** (4): 383–8.

Peluso, N. L. (1992) *Rich Forests, Poor People: Resource Control and Resistance in Java*. Berkeley, University of California Press.

Pelzer, K. (1978) "Swidden Cultivation in South East Asia: Historical, Ecological, and Economic Perspectives." In *Farmers in the Forest: Economic Development and Marginal Agriculture in Northern Thailand*, edited by P. Kunstadter. Honolulu, University of Hawaii Press.

Pento, T. (1999) "Industrial Ecology of the Paper Industry." *Water Science and Technology* **40** (11–12): 21–4.

Perreault, T. (2001) "Developing Identities: Indigenous Mobilization, Rural Livelihoods, and Resource Access in Ecuadorian Amazonia." *Ecumene* **8** (4): 381–413.

Petraitis, P. S., R. E. Latham and R. A. Niesenbaum (1989) "The Maintenance of Species Diversity by Disturbance." *The Quarterly Review of Biology* **64** (4): 393–416.

Poole, P. (1995) "Geomatics: Who Needs it?" *Cultural Survival Quarterly* **18** (4): 1.

Porter, P. W. and E. S. Sheppard (1998) *A World of Difference: Society, Nature, Development*. New York, Guilford Press.

Porter, T. M. (1995) *Trust in Numbers: The Pursuit of Objectivity in Science and Public Life*. Princeton, Princeton University Press.

Prasad, R. R. (1994) *Pastoral Nomadism in Arid Zones of India*. New Delhi, Discovery Publishing House.

Pred, A. and M. J. Watts (1992) *Reworking Modernity: Capitalisms and Symbolic Discontent*. New Brunswick (NJ), Rutgers University Press.

Proctor, J. D. (1998) "The Social Construction of Nature: Relativist Accusations, Pragmatist and Critical Realist Responses." *Annals of the Association of American Geographers* **88** (3): 352–76.

Pulido, L. (1996) *Environmentalism and Economic Justice: Two Chicano Struggles in the Southwest*. Tuscon, University of Arizona Press.

Raby, P. (2001) *Alfred Russel Wallace: A Life*. Princeton, Princeton University Press.

Rangan, H. (1997) "Indian Environmentalism and the Question of the State: Problems and Prospects for Sustainable Development." *Environment and Planning A* **29**: 2129–43.

Rangan, H. (2000) *Of Myths and Movements: Rewriting Chipko into Himalayan History*. London, Verso.

Rappaport, R. (1975) "The Flow of Energy in an Agricultural Society." In *Biological Anthropology*, edited by S. H. Katz. San Francisco, W. H. Freeman: 117–32.

Rappaport, R. A. (1967) "Ritual Regulation of Environmental Relations among a New Guinea People." *Ethnology* **6**: 17–30.

Rappaport, R. A. (1968) *Pigs for the Ancestors: Ritual in the Ecology of a New Guinea People*. New Haven, Yale University Press.

Reclus, E. (1871) *The Earth: A Descriptive History of the Phenomena of the Life of the Globe*. New York, Harper and Brothers.

Reclus, E. (1890) *Evolution and Revolution*. London, W. Reeves.

Rees, J. (1990) *Natural Resources: Allocation, Economics, and Policy*. New York, Routledge.

Reynolds, J. F. and M. Stafford Smith, eds. (2002) *Global Desertification: Do Humans Cause Deserts?* New York, John Wiley and Sons.

Ribot, J. C. (1996) "Participation without Representation: Chiefs, Councils, and Forestry Law in the West African Sahel." *Cultural Survival Quarterly* Fall: 40–44.

Richards, J. F. (1990) "Land Transformation." In *The Earth as Transformed by Human Action*, edited by B. L. T. Turner, W. C. Clark, R. W. Kates et al. Cambridge, Cambridge University Press: 163–78.

Robbins, P. (1998a) "Authority and Environment: Institutional Landscapes in Rajasthan, India." *Annals of the Association of American Geographers* **88** (3): 410–35.

Robbins, P. (1998b) "Paper Forests: Imagining and Deploying Exogenous Ecologies in Arid India." *Geoforum* **29** (1): 69–86.

Robbins, P. (2000) "The Practical Politics of Knowing: State Environmental Knowledge and Local Political Economy." *Economic Geography* **76** (2): 126–44.

Robbins, P. (2001) "Tracking Invasive Land Covers in India, or, Why our Landscapes have Never been Modern." *Annals of the Association of American Geographers* **91** (4): 637–59.

Robbins, P. (2002) "Poverty and Gender in Indian Food Security: Assessing Measures of Inequity." In *Food Security and Environmental Quality in the Developing World*, edited by R. Lal, D. O. Hansen, N. Uphoff and S. Slack. Boca Raton (US), CRC Press: 369–82.

Robbins, P. and T. Birkenholtz (2003) "Turfgrass Revolution: Measuring the Expansion of the American Lawn." *Land Use Policy* **20**: 181–94.

Robbins, P. and J. Sharp (2003) "Producing and Consuming Chemicals: The Moral Economy of the American Lawn." *Economic Geography* **79** (4): 425–51.

Robbins, P., A.-M. Polderman and T. Birkenholtz (2001) "Lawns and Toxins: An Ecology of the City." *Cities: The International Journal of Urban Policy and Planning* **18** (6): 369–80.

Rocheleau, D. and D. Edmunds (1997) "Women, Men and Trees: Gender, Power and Property in Forest and Agrarian Landscapes." *World Development* **25** (8): 1351–71.

Rocheleau, D. and L. Ross (1995) "Trees as Tools, Trees as Text: Struggles over Resources in Zambrana-Chacuey, Domincan Republic." *Antipode* **27** (4): 407.

Rocheleau, D., B. Thomas-Slayter and E. Wangari (1996) "Gender and Environment: A Feminist Political Ecology Perspective." In *Feminist Political Ecology: Global Issues and Local Experience*, edited by D. Rocheleau, B. Thomas-Slayter and E. Wangari. New York, Routledge: 3–23.

Rocheleau, D., L. Ross, J. Morrobel, L. Malaret, R. Hernandez and T. Kominiak (2001) "Complex Communities and Emergent Ecologies in the Regional Agroforest of Zambrana-Chacuey, Dominican Republic." *Ecumene* **8** (4): 465–92.

Roeder, P. G. (1984) "Legitimacy and Peasant Revolution: An Alternative to Moral Economy." *Peasant Studies* **11** (3): 149–68.

Rosin, R. T. (1993) "The Tradition of Groundwater Irrigation in Northwestern India." *Human Ecology* **21** (1): 51–86.

Rozanov, B. G., V. Targulian and D. S. Orlov (1990) "Soils." In *The Earth as Transformed by Human Action*, edited by B. L. T. Turner, W. C. Clark, R. W. Kates et al. Cambridge, Cambridge University Press: 203–14.

Russell, E. W. B. (1997) *People and the Land through Time: Linking Ecology and History*. New Haven, Conn., Yale University Press.

Said, E. (1978) *Orientalism*. New York, Random House.

Sainath, P. (1996) *Everybody Loves a Good Drought*. New York, Penguin Books.

St Martin, K. (2001) "Making Space for Community Resource Management in Fisheries." *Annals of the Association of American Geographers* **91** (1): 122–42.

Salter, L. (1988) *Mandated Science*. Boston, Kluwer Academic Publishers.

Sambrook, R. A., B. W. Pigozzi and R. N. Thomas (1999) "Population Pressure, Deforestation, and Land Degradation: A Case Study from the Dominican Republic." *Professional Geographer* **51** (1): 25–40.

Sauer, C. O. (1965a) "The Morphology of Landscape." In *Land and Life: A Selection from the Writings of Carl Ortwin Sauer*, edited by J. Leighly. Berkeley, University of California Press: 315–50.

Sauer, C. O. (1965b) "The Theme of Plant and Animal Destruction in Economic History." In *Land and Life: A Selection from the Writings of Carl Ortwin Sauer*, edited by J. Leighly. Berkeley, University of California Press: 145–54.

Saunders, P. L. (2000) "Environmental Refugees: The Origins of a Construct." In *Political Ecology: Science, Myth, and Power*, edited by P. Stott and S. Sullivan. New York, Arnold.

Schmink, M. and C. H. Wood (1987) "The 'Political Ecology' of Amazonia." In *Lands at Risk in the Third World: Local Level Perspectives*, edited by P. D. Little, M. H. Horowitz and A. E. Nyerges. Boulder, Westview Press: 38–57.

Schmink, M. and C. H. Wood (1992) *Contested Frontiers in Amazonia*. New York, Columbia University Press.

Schroeder, R. A. (1999) *Shady Practices: Agroforestry and Gender Politics in the Gambia*. Berkeley, University of California Press.

Schwarz, H. E., J. Emel, W. J. Dickens and P. Rogers (1990) "Soils." In *The Earth as Transformed by Human Action*, edited by B. L. T. Turner, W. C. Clark, R. W. Kates et al. Cambridge, Cambridge University Press: 253–70.

Scott, A. (1990) *Ideology and the New Social Movements*. London, Unwin Hyman.

Scott, J. C. (1976) *The Moral Economy of the Peasant: Rebellion and Subsistence in Southeast Asia*. New Haven, Yale University Press.

Scott, J. C. (1985a) "Socialism and Small Property, or, Two Cheers for the Petty Bourgeoisie." *Peasant Studies* **12** (3): 185–97.

Scott, J. C. (1985b) *Weapons of the Weak: Everyday Forms of Peasant Resistance*. New Haven, Yale University Press.

Scott, J. C. (1998) *Seeing Like a State: How Certain Schemes to Improve the Human Condition have Failed*. New Haven and London, Yale University Press.

Seager, J. (1996) "'Hysterical Housewives' and other Mad Women: Grassroots Environmental Organizing in the United States." In *Feminist Political Ecology: Global Issues and Local Experiences*, edited by D. Rocheleau, B. Thomas-Slayter and E. Wangari. New York, Routledge: 271–83.

Shankanarayan, K. A. and Y. Styanarayan (1964) "Grazing Resources of Rajasthan: Grassland Types of Alluvial Plains." *Indian Forester* **90**: 436–41.

Sharon, R. and H. Dayal (1993) "Deprivation of Female Farm Laborers in Jharkand Region of Bihar." *Social Change* **23** (4): 95–9.

Shiva, V. (1988) *Staying Alive: Women, Ecology and Development*. London, Zed Books.

Shiva, V. (1991) *The Violence of the Green Revolution*. Penang, Malayasia, Third World Network.

Simoons, F. J. (1979) "Questions in the Sacred Cow Controversy." *Current Anthropology* **20** (3): 467–76.

Singh, R. D. (1996) "Female Agricultural Workers' Wages, Male–Female Wage Differentials, and Agricultural Growth in a Developing Country, India." *Economic Development and Cultural Change* **45** (1): 89–123.

Sismondo, S. (1993) "Some Social Constructions." *Social Studies of Science* **23**: 515–53.

Sivaramakrishnan, K. (1996) "The Politics of Fire and Forest Regeneration in Colonial Bengal." *Environment and History* **2**: 145–94.

Sivaramakrishnan, K. (1998) "Comanaged Forests in West Bengal: Historical Perspectives on Community and Control." *Journal of Sustainable Forestry* **7** (3/4): 23–49.

Sivaramakrishnan, K. (2000) "State Sciences and Development Histories: Encoding Local Forestry Knowledge in Bengal." *Development and Change* **31** (1): 61–89.

Slater, C. (1996) "Amazonia as Edenic Narrative." In *Uncommon Ground: Rethinking the Human Place in Nature*, edited by W. Cronon. New York, W. W. Norton and Co.: 114–31.

Sluyter, A. (1999) "The Making of the Myth in Postcolonial Development: Material-Conceptual Landscape Transformation in Sixteenth Century Veracruz." *Annals of the Association of American Geographers* **89** (3): 377–401.

Smith, N. (1996) "The Production of Nature." In *FutureNatural: Nature/Science/Culture*, edited by G. Robertson, M. Mash, L. Tickner et al. New York, Routledge: 35–54.

Smyth, C. G. and S. A. Royle (2000) "Urban Landslide Hazards: Incidence and Causative Factors in Niteroi, Rio de Janeiro State, Brazil." *Applied Geography* **20** (2): 95–117.

Sommerville, M. (1848) *Physical Geography*. London, John Murray.

Soysa, P. (1987) "Women and Nutrition." In *World Review of Nutrition and Dietetics*, edited by G. H. Bourne. Basel (Switzerland), Karger: **52**.

Spencer, T. and S. Whatmore (2001) "Editorial: Bio-geographies: Putting Life Back into the Discipline." *Transactions of the Institute of British Geographers* **26** (2): 139–41.

Speth, W. W. (1978) "The Anthropogeographic Theory of Franz Boas." *Anthropos* **73**: 1–31.

Speth, W. W. (1981) "Berkeley Geography, 1923–33." In *The Origins of Academic Geography in the United States*, edited by B. W. Blouet. Hambden (CT), Archon Books: 221–44.

Spirn, A. W. (1996) "Constructing Nature: The Legacy of Frederick Law Olmsted." In *Uncommon Ground: Rethinking the Human Place in Nature*, edited by W. Cronon. New York, Norton.

Spivak, G. C. (1990) *The Post-colonial Critic: Interviews, Strategies, Dialogues*, edited by S. Harasym. New York, Routledge.

Spraos, J. (1983) *Inequalizing Trade*. Oxford, Clarendon Press.

State of Marwar (1887) *Report on the Administration of the Jodhpur State for the Year 1886–1887*. Jodhpur, Marwar State Press.

Steward, J. H. (1972) *Theory of Culture Change: The Methodology of Multilinear Evolution*. Urbana, University of Illinois Press.

Stock, R. (1995) *Africa South of the Sahara: A Geographical Interpretation*. New York, Guilford Press.

Stott, P. and S. Sullivan, eds. (2000) *Political Ecology: Science, Myth and Power*. London, Arnold.

Sullivan, S. (2000) "Getting the Science Right, or, Introducing Science in the First Place?" In *Political Ecology: Science, Myth, and Power*, edited by P. Stott and S. Sullivan. New York, Arnold: 15–44.

Szasz, A. (1994) *Ecopopulism: Toxic Waste and the Movement for Environmental Justice*. Minneapolis, University of Minnesota Press.

Thompson, E. P. (1978) *The Poverty of Theory and other Essays*. New York, Monthly Review Press.

Trimbur, T. J. and M. Watts (1976) "Are Cultural Ecologists Well Adapted? A Review of the Concept of Adaptation." *Proceedings of the Association of American Geographers* **8**: 179–83.

Trotsky, L. (1962) *The Permanent Revolution and Results and Prospects*. London, New Park Publications.

Turner, B. L. (1989) "The Specialist-Synthesis Approach to the Revival of Geography: The Case of Cultural Ecology." *Annals of the Association of American Geographers* **79** (1): 88–100.

Turner, B. L. (1990) "The Rise and Fall of Population and Agriculture in the Central Maya Lowlands: 300 BC to Present." In *Hunger and History: Food Shortage, Poverty, and Deprivation*, edited by L. Newman. Cambridge, Blackwell Publishers.

Turner, B. L. (2002) "Contested Identities: Human–Environment Geography and Disciplinary Implications in a Restructuring Academy." *Annals of the Association of American Geographers* **92** (1): 52–74.

Turner, B. L. and S. B. Brush, eds. (1987) *Comparative Farming Systems*. New York, Guilford Press.

Turner, B. L., S. C. Villar, D. Foster et al. (2001) "Deforestation in the Southern Yucatan Peninsular Region: An Integrative Approach." *Forest Ecology and Management* **154** (3): 353–70.

Turner, M. D. (1998) "The Interaction of Grazing History with Rainfall and its Influence on Annual Rangeland Dynamics in the Sahel." In *Nature's Geography: New Lessons for*

Conservation in Developing Countries, edited by K. S. Zimmerer and K. R. Young. Madison (WI), University of Wisconsin Press: 237–61.

Turner, M. D. (1999) "Conflict, Environmental Change, and Social Institutions in Dryland Africa: Limitations of the Community Resource Management Approach." *Society and Natural Resources* **12** (7): 643–57.

Ulmen, G. L. (1978) *The Science of Society: Toward an Understanding of the Life and Work of Karl August Wittfogel.* The Hague, Mouton Publishers.

Uphoff, N. (1988) "Assisted Self-reliance: Working with, Rather than for the Poor." In *Strengthening the Poor: What have we Learned?*, edited by J. P. Lewis. Washington, DC, Transaction Books.

Vandermeer, J. and I. Perfecto (1995) *Breakfast of Biodiversity: The Truth about Rainforest Destruction.* Oakland (CA), Food First.

Vayda, A. P. (1983) "Progressive Contextualization: Methods for Research in Human Ecology." *Human Ecology* **11** (3): 265–81.

Vayda, A. P. and B. B. Walters (1999) "Against Political Ecology." *Human Ecology* **27** (1): 167–79.

Wade, R. (1987) "The Management of Common Property Resources: Collective Action as an Alternative to Privatization or State Regulation." *Cambridge Journal of Economics* **11**: 95–106.

Walker, P. (2003) "Reconsidering 'Regional' Political Ecologies: Toward a Political Ecology of the Rural American West." *Progress in Human Geography* **27** (1): 7–24.

Walker, R., E. Moran and L. Anselin (2000) "Deforestation and Cattle Ranching in the Brazilian Amazon: External Capital and Household Processes." *World Development* **28** (4): 683–99.

Warren, A., S. Batterbury and H. Osbahr (2001) "Soil Erosion in the West African Sahel: A Review and an Application of a 'Local Political Ecology' Approach in South West Niger." *Global Environmental Change – Human and Policy Dimensions* **11** (1): 79–95.

Watts, M. J. (1983a) "On the Poverty of Theory: Natural Hazards Research in Context." In *Interpretations of Calamity*, edited by K. Hewitt. Boston, Allan and Unwin: 231–62.

Watts, M. J. (1983b) *Silent Violence: Food, Famine and Peasantry in Northern Nigeria.* Berkeley, University of California Press.

Watts, M. J. (2000) "Political Ecology." In *A Companion to Economic Geography*, edited by E. Sheppard and T. Barnes. Malden (MA), Blackwell Publishers: 257–74.

Weber, M. (1978) *Economy and Society.* Berkeley, University of California Press.

Westoby, M., B. Walker and I. Noy-Meir (1989) "Opportunistic Management for Rangelands not at Equilibrium." *Journal of Range Management* **42** (4): 266–74.

Whatmore, S. (2003) "From Banana Wars to Black Sigatoka: Another Case for a More-than-human Geography." *Geoforum* **34** (2): 139.

White, G. F. (1945) *Human Adjustments to Floods: A Geographical Approach to the Flood Problem in the United States.* Chicago, University of Chicago, Dept. of Geography, Research paper no. 29.

White, G. F. and J. E. Haas (1975) *Assessment of Research on Natural Hazards.* Cambridge (MA), MIT Press.

Whiteside, K. H. (2002) *Divided Natures: French Contributions to Political Ecology.* Cambridge (MA), MIT Press.

Willems-Braun, B. (1997) "Buried Epistemologies: The Politics of Nature in (Post)colonial British Columbia." *Annals of the Association of American Geographers* **87** (1): 3–31.

Winkelman, M. (1998) "Aztec Human Sacrifice: Cross-cultural Assessments of the Ecological Hypothesis." *Ethnology* **37** (3): 285–98.

Wittfogel, K. A. (1981) *Oriental Despotism: A Comparative Study of Total Power.* New York, Vintage Books.

Wolf, E. (1972) "Ownership and Political Ecology." *Anthropological Quarterly* **45**: 201–5.

Wolf, E. R. (1969) *Peasant Wars of the Twentieth Century.* New York, Harper and Row.

Woodcock, G. and I. Avakumovic (1990) *From Prince to Rebel.* Montreal, Black Rose Press.

Woolgar, S. (1988) *Science: The Very Idea.* London, Tavistock.

World Wildlife Fund (2003) "Global 2000 Blueprint for a Living Planet: Ecoregion 10: Madagascar Forests and Shrubland." http://www.panda.org/about_wwf/where_we_work/ecoregions/global200/pages/region010.htm (accessed December 16, 2003).

Worster, D. (1979) *Dust Bowl: The Southern High Plains in the 1930s.* Oxford, Oxford University Press.

Worster, D. (1985a) *Nature's Economy: A History of Ecological Ideas.* Cambridge, Cambridge University Press.

Worster, D. (1985b) *Rivers of Empire: Water, Aridity, and the Growth of the American West.* New York, Pantheon.

Wynn, G. (1997) "Remapping Tutira: Contours in the Environmental History of New Zealand." *Journal of Historical Geography* **23** (4).

Zhang, X., D. E. Walling, T. A. Quine and A. Wen (1997) "Use of Reservoir Deposits and Caesium-137 Measurements to Investigate the Erosional Response of a Small Drainage Basin in the Rolling Loess Plateau Region of China." *Land Degradation and Development* **8** (1): 1–16.

Zimmerer, K. (1993) "Soil Erosion and Social (Dis)courses in Cochambamba, Bolivia: Perceiving the Nature of Environmental Degradation." *Economic Geography* **69** (3): 312–27.

Zimmerer, K. S. (1991) "Wetland Production and Smallholder Persistence: Agricultural Change in a Highland Peruvian Region." *Annals of the Association of American Geographers* **81** (3): 443–63.

Zimmerer, K. S. (2000) "The Reworking of Conservation Geographies: Nonequilibrium Landscapes and Nature–Society Hybrids." *Annals of the Association of American Geographers* **90** (2): 356–69.

Zimmerer, K. S. and T. J. Bassett, eds. (2003) *Political Ecology: An Integrative Approach to Geography and Environment-Development Studies.* New York: Guilford Press.

Zurayk, R., F. el-Awar, S. Hamadeh et al. (2001) "Using Indigenous Knowledge in Land Use Investigations: A Participatory Study in a Semi-arid Mountainous Region of Lebanon." *Agriculture Ecosystems and Environment* **86** (3): 247–62.

Index